Linear Algebra for Beginners
Open Doors to Great Careers

Richard Han

Copyright © 2018 Richard Han

All rights reserved.

CONTENTS

PREFACE 7
1 - INTRODUCTION 8
2 – SOLVING SYSTEMS OF LINEAR EQUATIONS 10
    GAUSSIAN ELIMINATION 10
    GAUSSIAN ELIMINATION AND ROW ECHELON FORM 12
    PROBLEM SET: GAUSSIAN ELIMINATION 15
    SOLUTION SET: GAUSSIAN ELIMINATION 16
    ELEMENTARY ROW OPERATIONS 20
    ELEMENTARY ROW OPERATIONS: ADDITIONAL EXAMPLE 22
    PROBLEM SET: ELEMENTARY ROW OPERATIONS 24
    SOLUTION SET: ELEMENTARY ROW OPERATIONS 25
    SUMMARY: SOLVING SYSTEMS OF LINEAR EQUATIONS 28
3 – VECTORS 29
    VECTOR OPERATIONS AND LINEAR COMBINATIONS 29
    PROBLEM SET: VECTOR OPERATIONS AND LINEAR COMBINATIONS 31
    SOLUTION SET: VECTOR OPERATIONS AND LINEAR COMBINATIONS 33
    VECTOR EQUATIONS AND THE MATRIX EQUATION $Ax=b$ 34
    LINEAR INDEPENDENCE 35
    LINEAR INDEPENDENCE: EXAMPLE 1 35
    LINEAR INDEPENDENCE: EXAMPLE 2 37
    PROBLEM SET: LINEAR INDEPENDENCE 39
    SOLUTION SET: LINEAR INDEPENDENCE 40
    SUMMARY: VECTORS 42
4 – MATRIX OPERATIONS 43
    ADDITION AND SCALAR MULTIPLICATION 43
    MULTIPLICATION 44
    PROBLEM SET: MATRIX OPERATIONS 46
    SOLUTION SET: MATRIX OPERATIONS 47

SUMMARY: MATRIX OPERATIONS 48

5 – PROPERTIES OF MATRIX ADDITION AND SCALAR MULTIPLICATION 49
COMMUTATIVITY, ASSOCIATIVITY, AND DISTRIBUTIVITY 49

IDENTITIES, ADDITIVE INVERSES, MULTIPLICATIVE ASSOCIATIVITY AND DISTRIBUTIVITY .. 51

PROBLEM SET: PROPERTIES OF MATRIX OPERATIONS 53

SOLUTION SET: PROPERTIES OF MATRIX OPERATIONS 54

TRANSPOSE OF A MATRIX 57

PROBLEM SET: TRANSPOSE OF A MATRIX 58

SOLUTION SET: TRANSPOSE OF A MATRIX 59

SUMMARY: PROPERTIES OF MATRIX ADDITION AND SCALAR MULTIPLICATION 60

6 – THE INVERSE OF A MATRIX 61
INVERSE MATRIX 61

GAUSS-JORDAN ELIMINATION 62

GAUSS-JORDAN ELIMINATION: ADDITIONAL EXAMPLE 64

PROBLEM SET: INVERSE OF A MATRIX 65

SOLUTION SET: INVERSE OF A MATRIX 66

SUMMARY: INVERSE OF A MATRIX 70

7 – DETERMINANTS 71
DETERMINANT OF A 2 BY 2 MATRIX 71

COFACTOR EXPANSION 71

COFACTOR EXPANSION: ADDITIONAL EXAMPLES 73

PROBLEM SET: DETERMINANTS 75

SOLUTION SET: DETERMINANTS 76

SUMMARY: DETERMINANTS 78

8 – PROPERTIES OF DETERMINANTS 79
DETERMINANT OF A PRODUCT OF MATRICES AND OF A SCALAR MULTIPLE OF A MATRIX 79

DETERMINANTS AND INVERTIBILITY 80

DETERMINANT OF THE TRANSPOSE OF A MATRIX 81

PROBLEM SET: PROPERTIES OF DETERMINANTS 82

SOLUTION SET: PROPERTIES OF DETERMINANTS 83

SUMMARY: PROPERTIES OF DETERMINANTS 85

9 – VECTOR SPACES 86
VECTOR SPACE DEFINITION 86

VECTOR SPACE EXAMPLE 86

VECTOR SPACE: ADDITIONAL EXAMPLE 88

PROBLEM SET: VECTOR SPACES 91

SOLUTION SET: VECTOR SPACES 92

EXAMPLES OF SETS THAT ARE NOT VECTOR SPACES 96

PROBLEM SET: SETS THAT ARE NOT VECTOR SPACES 97

SOLUTION SET: SETS THAT ARE NOT VECTOR SPACES 98

SUMMARY: VECTOR SPACES 100

10 – SUBSPACES 101
SUBSPACE DEFINITION AND SUBSPACE PROPERTIES 101

DEFINITION OF TRIVIAL AND NONTRIVIAL SUBSPACE 102

ADDITIONAL EXAMPLE OF SUBSPACE 102

PROBLEM SET: SUBSPACES 103

SOLUTION SET: SUBSPACES 104

SUBSETS THAT ARE NOT SUBSPACES 105

SUBSETS THAT ARE NOT SUBSPACES: ADDITIONAL EXAMPLE 105

PROBLEM SET: SUBSETS THAT ARE NOT SUBSPACES 106

SOLUTION SET: SUBSETS THAT ARE NOT SUBSPACES 107

SUMMARY: SUBSPACES 108

11 – SPAN AND LINEAR INDEPENDENCE 109
SPAN 109

SPAN OF A SUBSET OF A VECTOR SPACE 111

LINEAR INDEPENDENCE 112

DETERMINING LINEAR INDEPENDENCE OR DEPENDENCE 113

PROBLEM SET: SPAN AND LINEAR INDEPENDENCE 116

SOLUTION SET: SPAN AND LINEAR INDEPENDENCE 117

SUMMARY: SPAN AND LINEAR INDEPENDENCE 120

12 – BASIS AND DIMENSION 121

BASIS 121

DIMENSION 123

PROBLEM SET: BASIS AND DIMENSION 126

SOLUTION SET: BASIS AND DIMENSION 127

COORDINATES 129

CHANGE OF BASIS 129

EXAMPLES OF FINDING TRANSITION MATRICES 130

PROBLEM SET: COORDINATES AND CHANGE OF BASIS 133

SOLUTION SET: COORDINATES AND CHANGE OF BASIS 134

SUMMARY: BASIS AND DIMENSION 136

CONCLUSION 137

INDEX 138

PREFACE

Welcome to Linear Algebra for Beginners: Open Doors to Great Careers. This is a first textbook in linear algebra. Be sure to get the companion online course Linear Algebra for Beginners here: https://www.onlinemathtraining.com/linear-algebra/. The online course can be very helpful in conjunction with this book.

The prerequisite for this book and the online course is a basic understanding of algebra.

I want you to succeed and prosper in your career, life, and future endeavors. I am here for you. Visit me at: https://www.onlinemathtraining.com/

1 - INTRODUCTION

Welcome to Linear Algebra for Beginners: Open Doors to Great Careers! My name is Richard Han. This is a first textbook in linear algebra.

Ideal student:

If you're a working professional needing a refresher on linear algebra or a complete beginner who needs to learn linear algebra for the first time, this book is for you. If your busy schedule doesn't allow you to go back to a traditional school, this book allows you to study on your own schedule and further your career goals without being left behind.

If you plan on taking linear algebra in college, this is a great way to get ahead.

If you're currently struggling with linear algebra or have struggled with it in the past, now is the time to master it.

Benefits of studying this book:

After reading this book, you will have refreshed your knowledge of linear algebra for your career so that you can earn a higher salary.

You will have a required prerequisite for lucrative career fields such as Data Science and Artificial Intelligence.

You will be in a better position to pursue a masters or PhD degree in machine learning and data science.

Why Linear Algebra is important:

- Famous uses of linear algebra include:
    - Computer graphics. Matrices are used to rotate figures in three-dimensional space.
    - Cryptography. Messages can be encrypted and decrypted using matrix operations.
    - Machine learning. Eigenvectors can be used to reduce the dimensionality of a data set, using a technique called Principal Component Analysis (PCA).

- Electrical networks. Electrical networks can be solved using systems of linear equations.
- Leontief Input-output model in economics. The necessary outputs of a list of industries can be found using matrix operations.
- Finance. Regression analysis can be used to estimate relationships between financial variables. For example, the relationship between the monthly return to a given stock and the monthly return to the S&P 500 can be estimated using a linear regression model. The model can, in turn, be used to forecast the future monthly return of the given stock.

What my book offers:

In this book, I cover core topics such as:

- **Gaussian Elimination**
- **Vectors**
- **Matrix Algebra**
- **Determinants**
- **Vector Spaces**
- **Subspaces**
- **Span and Linear Independence**
- **Basis and Dimension**

I explain each definition and go through each example step by step so that you understand each topic clearly. Throughout the book, there are practice problems for you to try. Detailed solutions are provided after each problem set.

I hope you benefit from the book.

Best regards,
Richard Han

2 – SOLVING SYSTEMS OF LINEAR EQUATIONS

GAUSSIAN ELIMINATION

In this section, we're going to look at solving systems of linear equations. We're going to look at the process of Gaussian elimination, and it has three things that you can do. The first thing you can do is switch two equations. The second thing is that you can multiply one equation by a nonzero number. The third thing that you can do is add a multiple of one equation to a second equation. This set of three things that you can do is called **Gaussian elimination**. This will make a lot more sense if we look at some examples. Let's say we had a system of equations like this:

$$x - 2y = 1$$
$$4x + y = 0$$

Here, you have two equations and two variables x and y. So this is a system of two equations in two variables. What we want to do here is try to get rid of the x variable. So let's do -4 times the first equation and add that to the second equation: $-4E_1 + E_2$. If we multiply the first equation by -4 and leave the second equation as it is, we get this:

$$-4x + 8y = -4$$
$$4x + y = 0$$

Adding these two equations, we get:

$$-4x + 8y = -4$$
$$4x + y = 0$$
$$\overline{}$$
$$0x + 9y = -4$$

$0x$ is just 0. So we get $9y = -4$. Let's solve for y. Divide both sides by 9. And you get $y = -\frac{4}{9}$.

I want to find what x is; so I'll plug the y-value back in to the first equation:

$$x - 2\left(-\frac{4}{9}\right) = 1$$

Now, simplifying this, we get:

$$x + \frac{8}{9} = 1$$

Solve for x by subtracting $\frac{8}{9}$ from both sides, and I get $x = \frac{1}{9}$.

So $x = \frac{1}{9}$ and $y = -\frac{4}{9}$. That's a solution to our original system of equations.

Let's do another example. Let's say we had this system of equations:
$$x - y = 1$$
$$2x - 2y = 3$$

Let's multiply the first equation by -2 and add to the second equation: $-2E_1 + E_2$.

We get:
$$-2x + 2y = -2$$
$$2x - 2y = 3$$

Adding these two equations, we get:
$$-2x + 2y = -2$$
$$2x - 2y = 3$$
$$\overline{}$$
$$0x + 0y = 1$$

The left-hand side equals to 0. So we get $0 = 1$, which is a contradiction. Since we get a contradiction, the original system of equations has no solution.

Let's do one more example. Suppose we have:
$$7x + 5y = 2$$
$$14x + 10y = 4$$

Let's try to get the coefficient of x in the first equation to be -14 so that the $x's$ cancel out when we add both equations. Let's do $-2E_1 + E_2$. We get:
$$-14x - 10y = -4$$
$$14x + 10y = 4$$

Adding the two equations, we get:
$$0x + 0y = 0$$

And so: $0 = 0$. That doesn't really tell me anything. If you look back at the original system of equations, notice the second equation is just twice the first equation. So, really, we only have just one equation, the first equation. The second equation is redundant. So, all we have is $7x + 5y = 2$. Note that y can be anything. Let y be some parameter t. Let's plug that in for y and solve for x:
$$7x + 5t = 2$$

Bring the $5t$ to the other side: $\qquad 7x = 2 - 5t$

Dividing both sides by 7: $\qquad x = \frac{2}{7} - \frac{5}{7}t$

So, the set of all solutions is going to be the set of all pairs $\left(\frac{2}{7} - \frac{5}{7}t, t\right)$ where t is any real number.

GAUSSIAN ELIMINATION AND ROW ECHELON FORM

For a system of 3 equations and 3 variables, we want to solve in a similar fashion by getting rid of the variables one by one until we have a triangular shape. Let's look at an example.

Suppose we have the following system of equations:

$$x + y + z = 0$$
$$-x + 2y + 3z = 1$$
$$3x - 3y + z = -1$$

We have three equations and three variables x, y, and z. Notice that, in the second equation, we have a $-x$; and, if we were to add that to the first equation, the x terms would cancel out. So let's take the first equation E_1 and add that to the second equation E_2. Replace the second equation with $E_1 + E_2$ like this:

$$x + y + z = 0$$
$$3y + 4z = 1$$
$$3x - 3y + z = -1$$

Now, let's try to get rid the x term in the third equation by multiplying the first equation by -3 and adding the result to the third equation like this: $-3E_1 + E_3$. Replace the third equation with $-3E_1 + E_3$ to get:

$$x + y + z = 0$$
$$3y + 4z = 1$$
$$-6y - 2z = -1$$

Let's try to get rid of the y term in the third equation by doing $2E_2 + E_3$ and replacing the third equation by the result:

$$x + y + z = 0$$
$$3y + 4z = 1$$
$$6z = 1$$

Look at the $6z$ in the third equation. We want the coefficient of z to be 1. So divide the third equation by 6: $\frac{1}{6}E_3$. Replace the third equation by the result to get:

$$x + y + z = 0$$
$$3y + 4z = 1$$
$$z = \frac{1}{6}$$

Look at the coefficient of y in the second equation. We want that to be 1. So let's divide the second equation by 3: $\frac{1}{3}E_2$. Replace the second equation by the result to get:

$$x + y + z = 0$$
$$y + \frac{4}{3}z = \frac{1}{3}$$
$$z = \frac{1}{6}$$

Now, notice that all the coefficients of the leading variables in each equation are 1. When you have a triangular shape like the above and all the leading coefficients are 1, then we say that the system of equations is in **row echelon form**.

Let's do another example. Suppose we have the following system of equations:

$$x - 2y + 5z = 2$$
$$3x + 2y - z = -2$$

Let's try to get rid of the x term in the second equation by performing: $-3E_1 + E_2$. Replace the second equation by the result to get:

$$x - 2y + 5z = 2$$
$$8y - 16z = -8$$

Looking at the second equation, notice that z can be anything. So let $z = t$, where t is a free variable. Plug in t for z in the second equation and solve for y:

$$8y - 16t = -8$$
$$8y = 16t - 8$$
$$y = 2t - 1$$

So we have y in terms of t. We have z and y in terms of t. Now, we want to solve for x. Let's use the first equation $x - 2y + 5z = 2$. Plug in what we got for y and z and solve for x:

$$x - 2(2t - 1) + 5t = 2$$
$$x - 4t + 2 + 5t = 2$$
$$x + 2 + t = 2$$
$$x + t = 0$$

$$x = -t$$

So, we have $x = -t, y = 2t - 1, z = t$. Any point $(-t, 2t - 1, t)$, where $t \in \mathbb{R}$, is a solution to the original system of equations.

Let's do one more example. Suppose we have the following system of equations:

$$x + y - z = 0$$
$$x - y + z = 1$$
$$2x + y - z = 0$$

If we take the first equation and subtract the second equation, we can get rid of the x-term. So let's do $E_1 - E_2$. Replace the second equation with the result:

$$x + y - z = 0$$
$$2y - 2z = -1$$
$$2x + y - z = 0$$

Now, look at the first equation and the third equation; we want to get rid of the x-term. So, let's do $-2E_1 + E_3$.

$$x + y - z = 0$$
$$2y - 2z = -1$$
$$-y + z = 0$$

Let's look at the second and third equations; let's get rid of the y-term. Perform $E_2 + 2E_3$.

$$x + y - z = 0$$
$$2y - 2z = -1$$
$$0 = -1$$

We arrive at a contradiction. Therefore, the original set of equations has no solution.

PROBLEM SET: GAUSSIAN ELIMINATION

Solve the system of linear equations using Gaussian elimination.

1. $\begin{aligned} x+2y&=0 \\ -x+y&=10 \end{aligned}$

2. $\begin{aligned} 3x-y&=3 \\ -4x+11y&=7 \end{aligned}$

3. $\begin{aligned} 8x-5y&=20 \\ -16x+10y&=-40 \end{aligned}$

4. $\begin{aligned} 2x+y&=13 \\ -4x-2y&=4 \end{aligned}$

5. $\begin{aligned} x-y-z&=1 \\ 2x+y+3z&=0 \\ 3x-y+z&=-1 \end{aligned}$

6. $\begin{aligned} x+2y+2z&=4 \\ -y+z&=-1 \\ x+y&=8 \end{aligned}$

7. $\begin{aligned} x-y-z&=3 \\ x-10y+10z&=0 \end{aligned}$

SOLUTION SET: GAUSSIAN ELIMINATION

1. $\begin{array}{l} x+2y=0 \\ -x+y=10 \end{array}$

Add Equation 1 to Equation 2 to get our new Equation 2.
$$x + 2y = 0$$
$$3y = 10$$

Solve the second equation for y to get $y = \frac{10}{3}$. Then plug this y value back into equation 1 to get $x + 2\left(\frac{10}{3}\right) = 0$. Solving for x gives $x = -\frac{20}{3}$.

2. $\begin{array}{l} 3x-y=3 \\ -4x+11y=7 \end{array}$

Take 4 times Equation 1 and add to 3 times Equation 2 to get our new Equation 2.
$$3x - y = 3$$
$$29y = 33$$

Solve for y in the second equation to get $y=\frac{33}{29}$. Plug this value into Equation 1 to solve for x. $3x - \frac{33}{29} = 3$.

$$3x = 3 + \frac{33}{29}$$
$$3x = \frac{120}{29}$$
$$x = \frac{40}{29}$$

3. $\begin{array}{l} 8x-5y=20 \\ -16x+10y=-40 \end{array}$

Take 2 times Equation 1 and add to Equation 2 to get the new second equation.
$$8x - 5y = 20$$
$$0 = 0$$

Now, y is a free variable. So let y=t, where t is a parameter. The first equation gives us $8x - 5t = 20$. Solve for x to get $x = \frac{5}{8}t + \frac{5}{2}$.

4. $\begin{array}{l} 2x+y=13 \\ -4x-2y=4 \end{array}$

Add 2 times Equation 1 to Equation 2 to get our new Equation 2.

$$2x + y = 13$$
$$0 = 30$$

Since the second equation is a contradiction, there is no solution.

5. $\begin{array}{l} x-y-z=1 \\ 2x+y+3z=0 \\ 3x-y+z=-1 \end{array}$

Take -2 times the first equation and add to the second equation to get our new second equation.

$$x - y - z = 1$$
$$3y + 5z = -2$$
$$3x - y + z = -1$$

Now, take -3 times the first equation and add to the third equation to get our new third equation.

$$x - y - z = 1$$
$$3y + 5z = -2$$
$$2y + 4z = -4$$

Now, take -2Eq.2+3Eq.3 to get our new Eq.3.

$$x - y - z = 1$$
$$3y + 5z = -2$$
$$2z = -8$$

Solve the third equation for z. z=-4. Plug this into the second equation and solve for y.

$$3y - 20 = -2$$
$$y = 6$$

Plug the z and y values into the first equation and solve for x.

$$x - 6 - (-4) = 1$$
$$x = 3$$

6. $\begin{array}{l} x+2y+2z=4 \\ -y+z=-1 \\ x+y=8 \end{array}$

Add the first equation to 2 times the second equation to get our new second equation.

$$x + 2y + 2z = 4$$
$$x + 4z = 2$$
$$x + y = 8$$

Take -2Eq.3+Eq.1 to get our new Eq.3.

$$x + 2y + 2z = 4$$
$$x + 4z = 2$$
$$-x + 2z = -12$$

Take Eq.2+Eq.3 to get our new Eq.3.

$$x + 2y + 2z = 4$$
$$x + 4z = 2$$
$$6z = -10$$

Now, solve for z in the third equation. $z = -\frac{5}{3}$.

Plug the z value into the second equation and solve for x.

$$x - \frac{20}{3} = 2$$
$$x = \frac{26}{3}$$

Plug the x and z values into the first equation and solve for y.

$$\frac{26}{3} + 2y - \frac{10}{3} = 4$$
$$2y + \frac{16}{3} = 4$$
$$y = -\frac{2}{3}$$

7. $x - y - z = 3$
    $x - 10y + 10z = 0$

Let's do Eq.1-Eq.2 to get our new Eq. 2.

$$x - y - z = 3$$
$$9y - 11z = 3$$

Now, z is a free variable, so let z=t. Then plug that value into z in the second equation.

$$9y - 11t = 3$$

Solve for y.

$$y = \frac{11}{9}t + \frac{1}{3}$$

Plug the values we got for y and z into the first equation and solve for x.

$$x - (\frac{11}{9}t + \frac{1}{3}) - t = 3$$
$$x - \frac{20}{9}t - \frac{1}{3} = 3$$
$$x = \frac{20}{9}t + \frac{10}{3}$$

ELEMENTARY ROW OPERATIONS

We can rewrite a system of equations using a matrix. For example, look at this system of equations:

$$x + y + z = 0$$

$$-x + 2y + 3z = 1$$

$$3x - 3y + z = -1$$

We can write a matrix that encapsulates this system of equations. We look at the coefficients of the variables in this system of equations. For the first equation, the coefficients of the variables are 1, 1, and 1. On the right hand side, we have the constant 0. So, for the first row of the matrix, we have:

$$\begin{bmatrix} 1 & 1 & 1 & 0 \end{bmatrix}$$

Now, move on to the second equation. The coefficients are -1, 2, and 3. The constant on the right hand side is 1. So fill in the second row for the matrix like so:

$$\begin{bmatrix} 1 & 1 & 1 & 0 \\ -1 & 2 & 3 & 1 \end{bmatrix}$$

Let's move on to the third equation. The coefficients are 3, -3, and 1. The constant is -1. So fill in the third row of the matrix like so:

$$\begin{bmatrix} 1 & 1 & 1 & 0 \\ -1 & 2 & 3 & 1 \\ 3 & -3 & 1 & -1 \end{bmatrix}$$

This matrix that we just formed is called the **augmented matrix**.

Now, we can solve the system of equations using the same three operations we used earlier. Instead of performing operations on equations, we can perform operations on rows. The operations are called **elementary row operations**. The first thing you can do is switch two rows. The second thing you can do is multiply one row by a nonzero number. The third thing you can do is add a multiple of one row to a second row. Here is the list of the three elementary row operations:

1. Switch two rows.
2. Multiply one row by a nonzero number.
3. Add a multiple of one row to a second row.

These are exactly the same three steps that you saw earlier in Gaussian elimination.

Let's look at the augmented matrix we had earlier.

$$\begin{bmatrix} 1 & 1 & 1 & 0 \\ -1 & 2 & 3 & 1 \\ 3 & -3 & 1 & -1 \end{bmatrix}$$

Let's look at the first two rows. If we were to add those two rows, the 1 and the -1 would cancel out. So let's add rows 1 and 2 to get a new row 2: $R1 + R2 \rightarrow R2$

$$\begin{bmatrix} 1 & 1 & 1 & 0 \\ 0 & 3 & 4 & 1 \\ 3 & -3 & 1 & -1 \end{bmatrix}$$

Look at the third row; we want to get rid of the 3. So do $-3R1 + R3 \rightarrow R3$

$$\begin{bmatrix} 1 & 1 & 1 & 0 \\ 0 & 3 & 4 & 1 \\ 0 & -6 & -2 & -1 \end{bmatrix}$$

Let's focus on the second and third rows; we want to get rid of the -6 in the third row. So let's do $2R2 + R3 \rightarrow R3$.

$$\begin{bmatrix} 1 & 1 & 1 & 0 \\ 0 & 3 & 4 & 1 \\ 0 & 0 & 6 & 1 \end{bmatrix}$$

Look at the third row; we want the leading coefficient 6 to be 1. So divide the third row by 6 to get a new third row: $\frac{1}{6}R3 \rightarrow R3$

$$\begin{bmatrix} 1 & 1 & 1 & 0 \\ 0 & 3 & 4 & 1 \\ 0 & 0 & 1 & \frac{1}{6} \end{bmatrix}$$

Look at the second row; we want the coefficient 3 to be 1. So perform $\frac{1}{3}R2 \rightarrow R2$.

$$\begin{bmatrix} 1 & 1 & 1 & 0 \\ 0 & 1 & 4/3 & 1/3 \\ 0 & 0 & 1 & \frac{1}{6} \end{bmatrix}$$

Notice the triangular shape of the matrix; all the leading coefficients in each row are 1. Furthermore, the leading coefficient of any row is to the right of the leading coefficient in the previous row. Also, any row of all zeroes is at the bottom of the matrix (in our example, there is no row of all zeroes). Since these three conditions of row-echelon form are satisfied, our matrix is in row-echelon form.

ELEMENTARY ROW OPERATIONS: ADDITIONAL EXAMPLE

Let's do an additional example. Suppose we had the following system of equations:

$$y + z = 0$$

$$x - y - z = 1$$

$$2x + 2y - z = 3$$

Let's write the corresponding augmented matrix. Note that, in the first equation, there is no x term; so the coefficient for x is 1. So the coefficients for the first equation are 0, 1, and 1. The augmented matrix so far looks like this:

$$\begin{bmatrix} 0 & 1 & 1 & 0 \\ & & & \\ & & & \end{bmatrix}$$

Filling in the second and third rows, we get:

$$\begin{bmatrix} 0 & 1 & 1 & 0 \\ 1 & -1 & -1 & 1 \\ 2 & 2 & -1 & 3 \end{bmatrix}$$

The first coefficient in the first row is 0, and we want a 1 there instead. So let's switch the first row with the second row: $R1 \leftrightarrow R2$

$$\begin{bmatrix} 1 & -1 & -1 & 1 \\ 0 & 1 & 1 & 0 \\ 2 & 2 & -1 & 3 \end{bmatrix}$$

Let's look at the third row. We want to get rid of the first 2 and make it a 0. So let's do $-2R1 + R3 \to R3$.

$$\begin{bmatrix} 1 & -1 & -1 & 1 \\ 0 & 1 & 1 & 0 \\ 0 & 4 & 1 & 1 \end{bmatrix}$$

Let's get rid of the 4 in the third row by doing $-4R2 + R3 \to R3$.

$$\begin{bmatrix} 1 & -1 & -1 & 1 \\ 0 & 1 & 1 & 0 \\ 0 & 0 & -3 & 1 \end{bmatrix}$$

Let's make the -3 in the third row a 1 by doing $-\frac{1}{3}R3 \to R3$.

$$\begin{bmatrix} 1 & -1 & -1 & 1 \\ 0 & 1 & 1 & 0 \\ 0 & 0 & 1 & -\dfrac{1}{3} \end{bmatrix}$$

Notice the triangular shape and that all the leading coefficients are 1. So this matrix is in row echelon form.

PROBLEM SET: ELEMENTARY ROW OPERATIONS

Solve the system of linear equations using an augmented matrix and elementary row operations.

1. $\begin{aligned} 7x+7y-z&=3 \\ x+y+z&=-1 \\ -x-y+3z&=0 \end{aligned}$

2. $\begin{aligned} 3x-y+11z&=7 \\ 7x+y+7z&=-3 \\ 14x+2y+14z&=-6 \end{aligned}$

3. $\begin{aligned} x-11y-z&=8 \\ 8x+y-z&=2 \\ -7x-12y&=-12 \end{aligned}$

4. $\begin{aligned} x-z&=2 \\ x+y+z&=-3 \\ x-y&=0 \end{aligned}$

5. $\begin{aligned} x+z&=8 \\ y+z&=-10 \end{aligned}$

6. $\begin{aligned} x-y+z&=9 \\ 9y-9z&=3 \end{aligned}$

SOLUTION SET: ELEMENTARY ROW OPERATIONS

1. $7x+7y-z=3$
   $x+y+z=-1$
   $-x-y+3z=0$

$$\begin{bmatrix} 7 & 7 & -1 & 3 \\ 1 & 1 & 1 & -1 \\ -1 & -1 & 3 & 0 \end{bmatrix} \quad -7R_2+R_1 \to R_2$$

$$\begin{bmatrix} 7 & 7 & -1 & 3 \\ 0 & 0 & -8 & 10 \\ -1 & -1 & 3 & 0 \end{bmatrix} \quad -7R_3+R_1 \to R_3$$

$$\begin{bmatrix} 7 & 7 & -1 & 3 \\ 0 & 0 & -8 & 10 \\ 0 & 0 & 20 & 3 \end{bmatrix} \quad -\frac{1}{8}R_2 \to R_2$$

$$\begin{bmatrix} 7 & 7 & -1 & 3 \\ 0 & 0 & 1 & -\frac{5}{4} \\ 0 & 0 & 20 & 3 \end{bmatrix} \quad -20R_2+R_3 \to R_3$$

$$\begin{bmatrix} 7 & 7 & -1 & 3 \\ 0 & 0 & 1 & -\frac{5}{4} \\ 0 & 0 & 0 & 28 \end{bmatrix}$$

The last row gives that 0=28, which is a contradiction. So there is no solution.

2. $3x-y+11z=7$
   $7x+y+7z=-3$
   $14x+2y+14z=-6$

$$\begin{bmatrix} 3 & -1 & 11 & 7 \\ 7 & 1 & 7 & -3 \\ 14 & 2 & 14 & -6 \end{bmatrix} \quad 7R_1-3R_2 \to R_2$$

$$\begin{bmatrix} 3 & -1 & 11 & 7 \\ 0 & -10 & 56 & 58 \\ 14 & 2 & 14 & -6 \end{bmatrix} \quad \frac{1}{2}R_3 \to R_3$$

$$\begin{bmatrix} 3 & -1 & 11 & 7 \\ 0 & -10 & 56 & 58 \\ 7 & 1 & 7 & -3 \end{bmatrix} \quad 7R_1-3R_3 \to R_3$$

$$\begin{bmatrix} 3 & -1 & 11 & 7 \\ 0 & -10 & 56 & 58 \\ 0 & -10 & 56 & 58 \end{bmatrix} \quad R_2-R_3 \to R_3$$

$$\begin{bmatrix} 3 & -1 & 11 & 7 \\ 0 & -10 & 56 & 58 \\ 0 & 0 & 0 & 0 \end{bmatrix}$$

From the second row, we have $-10y + 56z = 58$. Since z is a free variable, let z=t. Then $-10y + 56t = 58$. Solve for y.

$$y = \frac{28}{5}t - \frac{29}{5}$$

Now, the first row tells us
$3x - y + 11z = 7$. Plug in the values we got for y and z. Then solve for x.

$$3x - \left(\frac{28}{5}t - \frac{29}{5}\right) + 11t = 7$$
$$x = -\frac{9}{5}t + \frac{2}{5}$$

So $x = -\frac{9}{5}t + \frac{2}{5}, y = \frac{28}{5}t - \frac{29}{5}, z = t, t \in \mathbb{R}$.

3. $\begin{array}{c} x-11y-z=8 \\ 8x+y-z=2 \\ -7x-12y=-12 \end{array}$

$$\begin{bmatrix} 1 & -11 & -1 & 8 \\ 8 & 1 & -1 & 2 \\ -7 & -12 & 0 & -12 \end{bmatrix} \quad \text{-8R}_1\text{+R}_2\rightarrow\text{R}_2$$

$$\begin{bmatrix} 1 & -11 & -1 & 8 \\ 0 & 89 & 7 & -62 \\ -7 & -12 & 0 & -12 \end{bmatrix} \quad \text{7R}_1\text{+R}_3\rightarrow\text{R}_3$$

$$\begin{bmatrix} 1 & -11 & -1 & 8 \\ 0 & 89 & 7 & -62 \\ 0 & -89 & -7 & 44 \end{bmatrix} \quad \text{R}_2\text{+R}_3\rightarrow\text{R}_3$$

$$\begin{bmatrix} 1 & -11 & -1 & 8 \\ 0 & 89 & 7 & -62 \\ 0 & 0 & 0 & -18 \end{bmatrix}$$

The last row tells us that 0=-18, which is a contradiction. Therefore, there is no solution.

4. $\begin{array}{c} x-z=2 \\ x+y+z=-3 \\ x-y=0 \end{array}$

$$\begin{bmatrix} 1 & 0 & -1 & 2 \\ 1 & 1 & 1 & -3 \\ 1 & -1 & 0 & 0 \end{bmatrix} \quad \text{R}_1\text{-R}_2\rightarrow\text{R}_2$$

$$\begin{bmatrix} 1 & 0 & -1 & 2 \\ 0 & -1 & -2 & 5 \\ 1 & -1 & 0 & 0 \end{bmatrix} \quad R_1-R_3 \to R_3$$

$$\begin{bmatrix} 1 & 0 & -1 & 2 \\ 0 & -1 & -2 & 5 \\ 0 & 1 & -1 & 2 \end{bmatrix} \quad R_2+R_3 \to R_3$$

$$\begin{bmatrix} 1 & 0 & -1 & 2 \\ 0 & -1 & -2 & 5 \\ 0 & 0 & -3 & 7 \end{bmatrix}$$

The third row tells us -3z=7. So $z=-\frac{7}{3}$.

The second row tells us $-y-2z=5$.

$-y - 2\left(-\frac{7}{3}\right) = 5$.

$y = -\frac{1}{3}$.

The first row tells us x-z=2. So $x - \left(-\frac{7}{3}\right) = 2$.

$x = -\frac{1}{3}$

5. $\begin{matrix} x+z=8 \\ y+z=-10 \end{matrix}$

$$\begin{bmatrix} 1 & 0 & 1 & 8 \\ 0 & 1 & 1 & -10 \end{bmatrix}$$

The second row tells us y+z=-10. Z is a free variable, so let z=t. Then y+t=-10. Y=-t-10.

The first row tells us x+z=8. So x+t=8. X=-t+8.

$$x = -t+8, y = -t-10, z = t, t \in \mathbb{R}$$

6. $\begin{matrix} x-y+z=9 \\ 9y-9z=3 \end{matrix}$

$$\begin{bmatrix} 1 & -1 & 1 & 9 \\ 0 & 9 & -9 & 3 \end{bmatrix}$$

The second row tells us 9y-9z=3. Let z=t. Then $y = t + \frac{1}{3}$.

The first row tells us $x - y + z = 9$.

$$x - \left(t + \frac{1}{3}\right) + t = 9$$

$$x = \frac{28}{3}$$

$$x = \frac{28}{3}, y = t + \frac{1}{3}, z = t, t \in \mathbb{R}$$

SUMMARY: SOLVING SYSTEMS OF LINEAR EQUATIONS

- Gaussian elimination consists of the following three actions:
  1. Switch two equations.
  2. Multiply one equation by a nonzero number.
  3. Add a multiple of one equation to a second equation.

- Elementary row operations consist of the following three actions:
  1. Switch two rows.
  2. Multiply one row by a nonzero number.
  3. Add a multiple of one row to a second row.

- A matrix is in row echelon form if the following three conditions hold:
  1. Any row of all zeroes is at the bottom of the matrix.
  2. The leading coefficient of any row is to the right of the leading coefficient in the previous row.
  3. All leading coefficients are 1.

3 – VECTORS

VECTOR OPERATIONS AND LINEAR COMBINATIONS

We will now learn about vector operations and linear combinations. What is a vector? A vector is a list of real numbers. Let's look at an example:

$$v = \begin{bmatrix} 0 \\ -2 \end{bmatrix}$$

This is a vector in $\mathbb{R}^2$. $\mathbb{R}^2$ symbolizes all pairs of real numbers. Consider another example:

$$u = \begin{bmatrix} 0 \\ 3 \\ -22 \end{bmatrix}$$

Notice this vector has three entries; so this is a vector in $\mathbb{R}^3$.

Now, we want to be able to add two vectors. We can add two vectors u_1 and u_2 by adding their corresponding entries. For example, suppose $u_1 = \begin{bmatrix} 0 \\ 3 \\ -22 \end{bmatrix}$ and $u_2 = \begin{bmatrix} 7 \\ 2 \\ 5 \end{bmatrix}$. To add u_1 and u_2, form the vector that you get when you add the corresponding entries in each vector.

$$u_1 + u_2 = \begin{bmatrix} 7 \\ 5 \\ -17 \end{bmatrix}$$

So, if we want to add two vectors, we just add their corresponding entries. The two vectors need to have the same number of entries.

We can also multiply a vector by a scalar c. A scalar is just a real number. Let's do an example. Suppose $c = 8$ and $v = \begin{bmatrix} 1 \\ -1 \\ 2 \end{bmatrix}$. We want to find cv. To multiply a vector by a scalar, we just multiply each entry of the vector by the scalar.

$$cv = 8\begin{bmatrix}1\\-1\\2\end{bmatrix} = \begin{bmatrix}8\\-8\\16\end{bmatrix}$$

If we have a set of vectors $v_1, v_2, \ldots, v_k$ and scalars $c_1, c_2, \ldots, c_k$, then $c_1 v_1 + c_2 v_2 + \cdots + c_k v_k$ is called a ***linear combination*** of $v_1, v_2, \ldots, v_k$ and $c_1, \ldots, c_k$ are called ***weights***. Let's look at an example.

Suppose $v_1 = \begin{bmatrix}1\\1\\2\end{bmatrix}, v_2 = \begin{bmatrix}0\\-1\\0\end{bmatrix}, v_3 = \begin{bmatrix}2\\2\\2\end{bmatrix}$ and $c_1 = 1, c_2 = 3, c_3 = -5$. Then $c_1 v_1 + c_2 v_2 + c_3 v_3$ is a linear combination of v_1, v_2, v_3 with weights $1, 3, -5$. We can simplify the linear combination:

$$c_1 v_1 + c_2 v_2 + c_3 v_3 = 1\begin{bmatrix}1\\1\\2\end{bmatrix} + 3\begin{bmatrix}0\\-1\\0\end{bmatrix} + (-5)\begin{bmatrix}2\\2\\2\end{bmatrix}$$

$$= \begin{bmatrix}1\\1\\2\end{bmatrix} + \begin{bmatrix}0\\-3\\0\end{bmatrix} + \begin{bmatrix}-10\\-10\\-10\end{bmatrix}$$

$$= \begin{bmatrix}-9\\-12\\-8\end{bmatrix}$$

We can also multiply a vector x by a matrix A. For example, suppose $A = \begin{bmatrix}1 & 0 & 2\\3 & 1 & -1\\2 & 2 & 0\end{bmatrix}$ and $x = \begin{bmatrix}5\\2\\4\end{bmatrix}$. We can multiply the matrix A to the vector x as follows:

$$Ax = \begin{bmatrix}1 & 0 & 2\\3 & 1 & -1\\2 & 2 & 0\end{bmatrix}\begin{bmatrix}5\\2\\4\end{bmatrix}$$

To multiply this out, look at the first row of A. We multiply the first entry in the first row with the first entry of , multiply the second entry in the first row with the second entry of x, multiply the third entry of the first row with the third entry of x, and add the results to get the first entry.:

$$\begin{bmatrix}1 & 0 & 2\\3 & 1 & -1\\2 & 2 & 0\end{bmatrix}\begin{bmatrix}5\\2\\4\end{bmatrix} = \begin{bmatrix}1\cdot 5 + 0\cdot 2 + 2\cdot 4\\ \\ \end{bmatrix} = \begin{bmatrix}13\\ \\ \end{bmatrix}$$

For the second entry, we do the same thing but with the second row of A:

$$\begin{bmatrix}1 & 0 & 2\\3 & 1 & -1\\2 & 2 & 0\end{bmatrix}\begin{bmatrix}5\\2\\4\end{bmatrix} = \begin{bmatrix}1\cdot 5 + 0\cdot 2 + 2\cdot 4\\ 3\cdot 5 + 1\cdot 2 + (-1)\cdot 4\\ \end{bmatrix} = \begin{bmatrix}13\\13\\ \end{bmatrix}$$

For the third entry, we do the same thing but with the third row of A:

$$\begin{bmatrix}1 & 0 & 2\\3 & 1 & -1\\2 & 2 & 0\end{bmatrix}\begin{bmatrix}5\\2\\4\end{bmatrix} = \begin{bmatrix}1\cdot 5 + 0\cdot 2 + 2\cdot 4\\ 3\cdot 5 + 1\cdot 2 + (-1)\cdot 4\\ 2\cdot 5 + 2\cdot 2 + 0\cdot 4\end{bmatrix} = \begin{bmatrix}13\\13\\14\end{bmatrix}$$

PROBLEM SET: VECTOR OPERATIONS AND LINEAR COMBINATIONS

1. Add the two vectors.

   a. $u=\begin{bmatrix}1\\0\end{bmatrix}$ $\quad v=\begin{bmatrix}-2\\1\end{bmatrix}$

   b. $u=\begin{bmatrix}1\\1\end{bmatrix}$ $\quad v=\begin{bmatrix}-1\\3\end{bmatrix}$

   c. $u=\begin{bmatrix}1\\1\\1\end{bmatrix}$ $\quad v=\begin{bmatrix}2\\-3\\10\end{bmatrix}$

2. Multiply the vector by the given scalar.

   a. $u=\begin{bmatrix}-1\\1\end{bmatrix}$ $\quad c=8$

   b. $u=\begin{bmatrix}2\\2\end{bmatrix}$ $\quad c=-3$

   c. $u=\begin{bmatrix}-9\\9\\18\end{bmatrix}$ $\quad c=-1$

3. Simplify the linear combination of vectors $c_1v_1+c_2v_2+c_3v_3$.

   a. $v_1=\begin{bmatrix}0\\1\end{bmatrix}$ $\quad v_2=\begin{bmatrix}-3\\6\end{bmatrix}$ $\quad v_3=\begin{bmatrix}11\\9\end{bmatrix}$
   $c_1=-1$ $\quad c_2=3$ $\quad c_3=4$.

   b. $v_1=\begin{bmatrix}2\\2\end{bmatrix}$ $\quad v_2=\begin{bmatrix}-1\\2\end{bmatrix}$ $\quad v_3=\begin{bmatrix}7\\7\end{bmatrix}$
   $c_1=3$ $\quad c_2=-3$ $\quad c_3=0$.

   c. $v_1=\begin{bmatrix}0\\0\\1\end{bmatrix}$ $\quad v_2=\begin{bmatrix}1\\1\\2\end{bmatrix}$ $\quad v_3=\begin{bmatrix}9\\8\\4\end{bmatrix}$
   $c_1=1$ $\quad c_2=6$ $\quad c_3=-1$.

   d. $v_1=\begin{bmatrix}4\\-4\\4\end{bmatrix}$ $\quad v_2=\begin{bmatrix}1\\1\\0\end{bmatrix}$ $\quad v_3=\begin{bmatrix}0\\0\\0\end{bmatrix}$
   $c_1=0$ $\quad c_2=1$ $\quad c_3=8$.

4. Find A**v**.

   a. $v=\begin{bmatrix}9\\-9\end{bmatrix}$ $\quad A=\begin{bmatrix}-2&3\\4&2\end{bmatrix}$

   b. $v=\begin{bmatrix}8\\-4\end{bmatrix}$ $\quad A=\begin{bmatrix}1&0\\-1&0\end{bmatrix}$

c. $v = \begin{bmatrix} -1 \\ 0 \\ 1 \end{bmatrix}$ $A = \begin{bmatrix} 2 & 3 & 4 \\ 0 & 1 & 0 \\ -7 & 7 & 14 \end{bmatrix}$

d. $v = \begin{bmatrix} -1 \\ 0 \\ 2 \end{bmatrix}$ $A = \begin{bmatrix} 1 & 0 & 0 \\ 0 & -1 & 1 \\ 1 & 1 & 2 \end{bmatrix}$

SOLUTION SET: VECTOR OPERATIONS AND LINEAR COMBINATIONS

1. a. $u + v = \begin{bmatrix} -1 \\ 1 \end{bmatrix}$

   b. $u + v = \begin{bmatrix} 0 \\ 4 \end{bmatrix}$

   c. $u + v = \begin{bmatrix} 3 \\ -2 \\ 11 \end{bmatrix}$

2. a. $cu = \begin{bmatrix} -8 \\ 8 \end{bmatrix}$

   b. $cu = \begin{bmatrix} -6 \\ -6 \end{bmatrix}$

   c. $cu = \begin{bmatrix} 9 \\ -9 \\ -18 \end{bmatrix}$

3. a. $c_1 v_1 + c_2 v_2 + c_3 v_3 = \begin{bmatrix} 35 \\ 53 \end{bmatrix}$

   b. $c_1 v_1 + c_2 v_2 + c_3 v_3 = \begin{bmatrix} 9 \\ 0 \end{bmatrix}$

   c. $c_1 v_1 + c_2 v_2 + c_3 v_3 = \begin{bmatrix} -3 \\ -2 \\ 9 \end{bmatrix}$

   d. $c_1 v_1 + c_2 v_2 + c_3 v_3 = \begin{bmatrix} 1 \\ 1 \\ 0 \end{bmatrix}$

4. a. $Av = \begin{bmatrix} -45 \\ 18 \end{bmatrix}$

   b. $Av = \begin{bmatrix} 8 \\ -8 \end{bmatrix}$

   c. $Av = \begin{bmatrix} 2 \\ 0 \\ 21 \end{bmatrix}$

   d. $Av = \begin{bmatrix} -1 \\ 2 \\ 3 \end{bmatrix}$

VECTOR EQUATIONS AND THE MATRIX EQUATION Ax=b

We're now ready to look at vector equations and the matrix equation of the form $Ax = b$. Recall the system of equations

$$x + y + z = 0$$

$$-x + 2y + 3z = 1$$

$$3x - 3y + z = -1$$

Look at the x variables in the system of equations and think of the x terms as one column. Similarly, think of the y terms as one column, think of the z terms as one column, and think of the constant terms on the right hand sides as one column. We can rewrite the system of equations as:

$$\begin{bmatrix} x \\ -x \\ 3x \end{bmatrix} + \begin{bmatrix} y \\ 2y \\ -3y \end{bmatrix} + \begin{bmatrix} z \\ 3z \\ z \end{bmatrix} = \begin{bmatrix} 0 \\ 1 \\ -1 \end{bmatrix}$$

Now, factor out the x from the first column, the y from the second column, and the z from the third column like this:

$$x \begin{bmatrix} 1 \\ -1 \\ 3 \end{bmatrix} + y \begin{bmatrix} 1 \\ 2 \\ -3 \end{bmatrix} + z \begin{bmatrix} 1 \\ 3 \\ 1 \end{bmatrix} = \begin{bmatrix} 0 \\ 1 \\ -1 \end{bmatrix}$$

So to ask if there is a solution to the system of equations is the same as asking if we can write $\begin{bmatrix} 0 \\ 1 \\ -1 \end{bmatrix}$ as a linear combination of the column vectors $\begin{bmatrix} 1 \\ -1 \\ 3 \end{bmatrix}, \begin{bmatrix} 1 \\ 2 \\ -3 \end{bmatrix}, \begin{bmatrix} 1 \\ 3 \\ 1 \end{bmatrix}$. In this case, x, y, z are the weights of the linear combination.

Now, let's define the span. The span of a set of vectors is the set of all linear combinations of those vectors. Thus, we want to know if the vector $\begin{bmatrix} 0 \\ 1 \\ -1 \end{bmatrix}$ lies in the $span \left\{ \begin{bmatrix} 1 \\ -1 \\ 3 \end{bmatrix}, \begin{bmatrix} 1 \\ 2 \\ -3 \end{bmatrix}, \begin{bmatrix} 1 \\ 3 \\ 1 \end{bmatrix} \right\}$. Note that the vector equation

$$x \begin{bmatrix} 1 \\ -1 \\ 3 \end{bmatrix} + y \begin{bmatrix} 1 \\ 2 \\ -3 \end{bmatrix} + z \begin{bmatrix} 1 \\ 3 \\ 1 \end{bmatrix} = \begin{bmatrix} 0 \\ 1 \\ -1 \end{bmatrix}$$

can be written as a matrix equation

$$\begin{bmatrix} 1 & 1 & 1 \\ -1 & 2 & 3 \\ 3 & -3 & 1 \end{bmatrix} \begin{bmatrix} x \\ y \\ z \end{bmatrix} = \begin{bmatrix} 0 \\ 1 \\ -1 \end{bmatrix}.$$

To see this, multiply out the left hand side to get:

$$\begin{bmatrix} x + y + z \\ -x + 2y + 3z \\ 3x - 3y + z \end{bmatrix}$$

We can rewrite this as the sum of the column vectors consisting of the variables:

$$\begin{bmatrix} x \\ -x \\ 3x \end{bmatrix} + \begin{bmatrix} y \\ 2y \\ -3y \end{bmatrix} + \begin{bmatrix} z \\ 3z \\ z \end{bmatrix}$$

Factoring out the variables, we get:

$$x\begin{bmatrix} 1 \\ -1 \\ 3 \end{bmatrix} + y\begin{bmatrix} 1 \\ 2 \\ -3 \end{bmatrix} + z\begin{bmatrix} 1 \\ 3 \\ 1 \end{bmatrix}$$

So, we get the linear combination of the coefficient vectors we had earlier. So our original system of equations can be rewritten as a matrix equation $Ax = b$ where A is the coefficient matrix, x is the column vector of weights $\begin{bmatrix} x \\ y \\ z \end{bmatrix}$, and b is our column vector of constants $\begin{bmatrix} 0 \\ 1 \\ -1 \end{bmatrix}$.

LINEAR INDEPENDENCE

We will now introduce the notion of linear independence. Let $v_1, \ldots, v_k \in \mathbb{R}^n$. Then the set $\{v_1, \ldots, v_k\}$ is **linearly independent** just in case the vector equation $c_1 v_1 + \cdots + c_k v_k = 0$ has only the trivial solution $c_1 = c_2 = \cdots = c_k = 0$. Otherwise, the set is said to be **linearly dependent**. Note the vector equation $c_1 v_1 + \cdots + c_k v_k = 0$ can be rewritten as $Ax = 0$, where $A = [v_1 \ldots v_k]$ and $x = \begin{bmatrix} c_1 \\ \vdots \\ c_k \end{bmatrix}$. In A, the vector v_1 forms the first column, v_2 forms the second column, etc.

LINEAR INDEPENDENCE: EXAMPLE 1

Let's do an example.

Determine if the set $\{v_1, v_2, v_3\}$ is linearly independent.

$$v_1 = \begin{bmatrix} 1 \\ -1 \\ 0 \end{bmatrix}, v_2 = \begin{bmatrix} 3 \\ 4 \\ 7 \end{bmatrix}, v_3 = \begin{bmatrix} 13 \\ -6 \\ 7 \end{bmatrix}$$

Form the matrix with v_1, v_2, v_3 as columns:

$$\begin{bmatrix} 1 & 3 & 13 \\ -1 & 4 & -6 \\ 0 & 7 & 7 \end{bmatrix}$$

Now, set an arbitrary linear combination of the vectors v_1, v_2, v_3 equal to 0 in matrix form:

$$\begin{bmatrix} 1 & 3 & 13 \\ -1 & 4 & -6 \\ 0 & 7 & 7 \end{bmatrix} \begin{bmatrix} c_1 \\ c_2 \\ c_3 \end{bmatrix} = \begin{bmatrix} 0 \\ 0 \\ 0 \end{bmatrix}$$

We want to know if this system of equations has only the trivial solution (in other words, $c_1 = c_2 = c_3 = 0$). Let's form the augmented matrix:

$$\begin{bmatrix} 1 & 3 & 13 & 0 \\ -1 & 4 & -6 & 0 \\ 0 & 7 & 7 & 0 \end{bmatrix}$$

Let's try to do a bunch of row operations on this augmented matrix to solve for the solution. If we get the trivial solution, then we know that the trivial solution would be the only solution, and the original set of vectors would be linearly independent. If we find that this system of equations has nontrivial solutions, then we know that the original set of vectors is linearly dependent.

Looking at the first two rows, let's do $R1 + R2 \to R2$:

$$\begin{bmatrix} 1 & 3 & 13 & 0 \\ 0 & 7 & 7 & 0 \\ 0 & 7 & 7 & 0 \end{bmatrix}$$

Look at the second and third rows. Let's do $R2 - R3 \to R3$:

$$\begin{bmatrix} 1 & 3 & 13 & 0 \\ 0 & 7 & 7 & 0 \\ 0 & 0 & 0 & 0 \end{bmatrix}$$

Looking at the leading coefficient 7 in the second row, let's make that a 1 by doing $\frac{1}{7}R2 \to R2$:

$$\begin{bmatrix} 1 & 3 & 13 & 0 \\ 0 & 1 & 1 & 0 \\ 0 & 0 & 0 & 0 \end{bmatrix}$$

Looking at the second row, notice that the value for the variable c_3 is free; it can be anything. Let $c_3 = t$, a free parameter. From the second row, we know that $c_2 + c_3 = 0$. Plugging in $c_3 = t$, solve for c_2:

$$c_2 + t = 0$$

$$c_2 = -t$$

From the first row, we know that $c_1 + 3c_2 + 13c_3 = 0$. Plugging in $c_2 = -t$ and $c_3 = t$, we get

$$c_1 + 3(-t) + 13t = 0$$

Solving for c_1, we get:

$$c_1 = -10t$$

Now we have $\begin{bmatrix} c_1 \\ c_2 \\ c_3 \end{bmatrix} = \begin{bmatrix} -10t \\ -t \\ t \end{bmatrix} = t \begin{bmatrix} -10 \\ -1 \\ 1 \end{bmatrix}$. Our set of solutions consists of all vectors of the form $t \begin{bmatrix} -10 \\ -1 \\ 1 \end{bmatrix}$. So there are many solutions besides the trivial solution. Since the system has a nontrivial solution, the set $\{v_1, v_2, v_3\}$ is linearly dependent. For example, let $t = 1$. Then, $c_1 = -10, c_2 = -1, c_3 = 1$. Remember that the matrix equation

$$\begin{bmatrix} 1 & 3 & 13 \\ -1 & 4 & -6 \\ 0 & 7 & 7 \end{bmatrix} \begin{bmatrix} c_1 \\ c_2 \\ c_3 \end{bmatrix} = \begin{bmatrix} 0 \\ 0 \\ 0 \end{bmatrix}$$

is equivalent to the vector equation $c_1 v_1 + c_2 v_2 + c_3 v_3 = 0$. Plugging in the values for c_1, c_2, c_3, we get $-10 v_1 - v_2 + v_3 = 0$. Therefore, we have a linear dependence relation between the vectors v_1, v_2, v_3.

LINEAR INDEPENDENCE: EXAMPLE 2

Let's look at a second example.

Determine if the set $\{v_1, v_2, v_3\}$ is linearly independent.

$$v_1 = \begin{bmatrix} 1 \\ 0 \\ -1 \end{bmatrix}, v_2 = \begin{bmatrix} 3 \\ 2 \\ 0 \end{bmatrix}, v_3 = \begin{bmatrix} 4 \\ -4 \\ 4 \end{bmatrix}$$

Form the augmented matrix with v_1, v_2, v_3 as columns:

$$\begin{bmatrix} 1 & 3 & 4 & 0 \\ 0 & 2 & -4 & 0 \\ -1 & 0 & 4 & 0 \end{bmatrix}$$

Let's try to solve this augmented matrix using row operations. Looking at the first and third rows, we can cancel out the leading terms, so let's do $R1 + R3 \to R3$:

$$\begin{bmatrix} 1 & 3 & 4 & 0 \\ 0 & 2 & -4 & 0 \\ 0 & 3 & 8 & 0 \end{bmatrix}$$

Looking at the second row, notice the leading coefficient is 2. Let's make that a 1 by doing $\frac{1}{2} R2 \to R2$:

$$\begin{bmatrix} 1 & 3 & 4 & 0 \\ 0 & 1 & -2 & 0 \\ 0 & 3 & 8 & 0 \end{bmatrix}$$

Let's get rid of the leading coefficient in the third row by doing $-3R2 + R3 \to R3$:

$$\begin{bmatrix} 1 & 3 & 4 & 0 \\ 0 & 1 & -2 & 0 \\ 0 & 0 & 14 & 0 \end{bmatrix}$$

From the third row, we know $14c_3 = 0$. So $c_3 = 0$. From the second row, we know $c_2 - 2c_3 = 0$. But $c_3 = 0$. So $c_2 - 2(0) = 0$. Therefore, $c_2 = 0$. From the first row, we know $c_1 + 3c_2 + 4c_3 = 0$. But we know $c_2 = c_3 = 0$. So $c_1 = 0$. Since $c_1 = c_2 = c_3 = 0$, the set of vectors v_1, v_2, v_3 is linearly independent.

PROBLEM SET: LINEAR INDEPENDENCE

1. Determine if the set $\{v_1, v_2, v_3\}$ is linearly independent.

   a. $v_1 = \begin{bmatrix} 6 \\ 6 \\ 7 \end{bmatrix}$ $v_2 = \begin{bmatrix} -1 \\ 0 \\ 8 \end{bmatrix}$ $v_3 = \begin{bmatrix} 2 \\ 2 \\ -2 \end{bmatrix}$

   b. $v_1 = \begin{bmatrix} 1 \\ 1 \\ 1 \end{bmatrix}$ $v_2 = \begin{bmatrix} 0 \\ 0 \\ 1 \end{bmatrix}$ $v_3 = \begin{bmatrix} 1 \\ 1 \\ 0 \end{bmatrix}$

   c. $v_1 = \begin{bmatrix} 1 \\ -1 \\ 0 \end{bmatrix}$ $v_2 = \begin{bmatrix} 2 \\ -2 \\ 0 \end{bmatrix}$ $v_3 = \begin{bmatrix} 7 \\ 7 \\ 9 \end{bmatrix}$

SOLUTION SET: LINEAR INDEPENDENCE

1. Determine if the set $\{v_1, v_2, v_3\}$ is linearly independent.

   a. $v_1 = \begin{bmatrix} 6 \\ 6 \\ 7 \end{bmatrix}$ $v_2 = \begin{bmatrix} -1 \\ 0 \\ 8 \end{bmatrix}$ $v_3 = \begin{bmatrix} 2 \\ 2 \\ -2 \end{bmatrix}$

   Form the matrix with the given vectors as columns and the zero column on the right.

   $\begin{bmatrix} 6 & -1 & 2 & 0 \\ 6 & 0 & 2 & 0 \\ 7 & 8 & -2 & 0 \end{bmatrix}$ $R_1 - R_2 \to R_2$

   $\begin{bmatrix} 6 & -1 & 2 & 0 \\ 0 & -1 & 0 & 0 \\ 7 & 8 & -2 & 0 \end{bmatrix}$ $7R_1 - 6R_3 \to R_3$

   $\begin{bmatrix} 6 & -1 & 2 & 0 \\ 0 & -1 & 0 & 0 \\ 0 & -55 & 26 & 0 \end{bmatrix}$ $-55R_2 + R_3 \to R_3$

   $\begin{bmatrix} 6 & -1 & 2 & 0 \\ 0 & -1 & 0 & 0 \\ 0 & 0 & 26 & 0 \end{bmatrix}$

   The third row tells us $26z=0$. So $z=0$. The second row tells us $-y=0$. So $y=0$. The first row tells us $6x-y+2z=0$. So $x=0$. The system of equations has only the trivial solution. Therefore, the given vectors are linearly independent.

   b. $v_1 = \begin{bmatrix} 1 \\ 1 \\ 1 \end{bmatrix}$ $v_2 = \begin{bmatrix} 0 \\ 0 \\ 1 \end{bmatrix}$ $v_3 = \begin{bmatrix} 1 \\ 1 \\ 0 \end{bmatrix}$

   Form the matrix with the given vectors as columns and the zero column on the right.

   $\begin{bmatrix} 1 & 0 & 1 & 0 \\ 1 & 0 & 1 & 0 \\ 1 & 1 & 0 & 0 \end{bmatrix}$ $R_1 - R_2 \to R_2$

   $\begin{bmatrix} 1 & 0 & 1 & 0 \\ 0 & 0 & 0 & 0 \\ 1 & 1 & 0 & 0 \end{bmatrix}$ $R_1 - R_3 \to R_3$

   $\begin{bmatrix} 1 & 0 & 1 & 0 \\ 0 & 0 & 0 & 0 \\ 0 & -1 & 1 & 0 \end{bmatrix}$ $R_2 \leftrightarrow R_3$

   $\begin{bmatrix} 1 & 0 & 1 & 0 \\ 0 & -1 & 1 & 0 \\ 0 & 0 & 0 & 0 \end{bmatrix}$

   The second row tells us $-y+z=0$. Let $z=t$. Then $y=t$. The first row tells us $x+z=0$. So $x=-t$.

All solutions will of the form $\begin{bmatrix} -t \\ t \\ t \end{bmatrix} = t \begin{bmatrix} -1 \\ 1 \\ 1 \end{bmatrix}$, where t is any real number. Since the system has non-trivial solutions, the given vectors are linearly dependent.

c. $v_1 = \begin{bmatrix} 1 \\ -1 \\ 0 \end{bmatrix}$ $v_2 = \begin{bmatrix} 2 \\ -2 \\ 0 \end{bmatrix}$ $v_3 = \begin{bmatrix} 7 \\ 7 \\ 9 \end{bmatrix}$

Form the matrix with the given vectors as columns and the zero column on the right.

$$\begin{bmatrix} 1 & 2 & 7 & 0 \\ -1 & -2 & 7 & 0 \\ 0 & 0 & 9 & 0 \end{bmatrix} \quad R_1 + R_2 \to R_2$$

$$\begin{bmatrix} 1 & 2 & 7 & 0 \\ 0 & 0 & 14 & 0 \\ 0 & 0 & 9 & 0 \end{bmatrix}$$

The second and third rows tell us that z=0. The first row tells us x+2y+7z=0. So x+2y=0. Letting y=t, we find x=-2t. So all solutions will be of the form $\begin{bmatrix} -2t \\ t \\ 0 \end{bmatrix} = t \begin{bmatrix} -2 \\ 1 \\ 0 \end{bmatrix}$. There are non-trivial solutions; so the given set of vectors are linearly dependent.

SUMMARY: VECTORS

- We add two vectors by adding their corresponding entries. We multiply a vector by a scalar by multiplying each entry in the vector by the scalar. We multiply matrix to a vector by taking the first row of the matrix and multiplying each entry in the first row by the corresponding entry in the vector, taking the second row of the matrix and multiplying each entry in the second row by the corresponding entry in the vector, etc.

- If we have a set of vectors $v_1, v_2, \ldots, v_k$ and scalars $c_1, c_2, \ldots, c_k$, then $c_1 v_1 + c_2 v_2 + \cdots + c_k v_k$ is called a *linear combination* of $v_1, v_2, \ldots, v_k$ and $c_1, \ldots, c_k$ are called *weights*.

- Let $v_1, \ldots, v_k \in \mathbb{R}^n$. Then the set $\{v_1, \ldots, v_k\}$ is *linearly independent* just in case the vector equation $c_1 v_1 + \cdots + c_k v_k = 0$ has only the trivial solution $c_1 = c_2 = \cdots = c_k = 0$. Otherwise, the set is said to be *linearly dependent*..

4 – MATRIX OPERATIONS

ADDITION AND SCALAR MULTIPLICATION

In this section, we're going to learn about matrix operations. But first, we want to know what a matrix is. An $m \times n$ matrix is an array $\begin{bmatrix} a_{11} & a_{12} & \cdots & a_{1n} \\ a_{21} & a_{22} & \cdots & a_{2n} \\ \vdots & & & \\ a_{m1} & a_{m2} & \cdots & a_{mn} \end{bmatrix}$ with m rows and n columns.

We can add two matrices if they have the same size. For example, suppose we want to add the following two matrices:

$$\begin{bmatrix} 2 & -1 & 0 \\ 1 & 1 & -1 \end{bmatrix} + \begin{bmatrix} 0 & 0 & 8 \\ 8 & 2 & 3 \end{bmatrix}$$

To add these two matrices, we just add their corresponding entries like this:

$$\begin{bmatrix} 2 & -1 & 0 \\ 1 & 1 & -1 \end{bmatrix} + \begin{bmatrix} 0 & 0 & 8 \\ 8 & 2 & 3 \end{bmatrix} = \begin{bmatrix} 2 & -1 & 8 \\ 9 & 3 & 2 \end{bmatrix}$$

We can also multiply a matrix by a scalar. For example, let's say the scalar is $c = 3$ and the matrix is $A = \begin{bmatrix} -1 & 2 & 4 \\ 0 & 1 & 1 \end{bmatrix}$. To find cA, we multiply each entry of A by the scalar c:

$$cA = 3 \begin{bmatrix} -1 & 2 & 4 \\ 0 & 1 & 1 \end{bmatrix} = \begin{bmatrix} -3 & 6 & 12 \\ 0 & 3 & 3 \end{bmatrix}$$

Let's do another example. Suppose $c = -2$ and $A = \begin{bmatrix} 11 & 10 \\ 8 & 9 \\ 4 & 2 \end{bmatrix}$.

$$cA = -2 \begin{bmatrix} 11 & 10 \\ 8 & 9 \\ 4 & 2 \end{bmatrix} = \begin{bmatrix} -22 & -20 \\ -16 & -18 \\ -8 & -4 \end{bmatrix}$$

We can define subtraction of two matrices $A - B$ as $A + (-1)B$. For example, suppose $A = \begin{bmatrix} 0 & 1 \\ -1 & 0 \end{bmatrix}$ and $B = \begin{bmatrix} 3 & 2 \\ 1 & 4 \end{bmatrix}$. Then we can find $A - B$ as follows:

$$A - B = A + (-1)B = \begin{bmatrix} 0 & 1 \\ -1 & 0 \end{bmatrix} + (-1) \begin{bmatrix} 3 & 2 \\ 1 & 4 \end{bmatrix}$$

$$= \begin{bmatrix} 0 & 1 \\ -1 & 0 \end{bmatrix} + \begin{bmatrix} -3 & -2 \\ -1 & -4 \end{bmatrix}$$

$$= \begin{bmatrix} -3 & -1 \\ -2 & -4 \end{bmatrix}$$

MULTIPLICATION

We can also multiply two matrices A and B as long as the number of columns of A is equal to the number of rows of B. Let's do an example. Suppose A is a 2×3 matrix and B is a 3×2 matrix as follows:

$$A = \begin{bmatrix} 1 & 3 & -1 \\ 0 & 2 & 2 \end{bmatrix} \text{ and } B = \begin{bmatrix} 1 & 0 \\ 1 & 0 \\ -1 & 1 \end{bmatrix}$$

The product AB will be a 2×2 matrix.

$$AB = \begin{bmatrix} 1 & 3 & -1 \\ 0 & 2 & 2 \end{bmatrix} \begin{bmatrix} 1 & 0 \\ 1 & 0 \\ -1 & 1 \end{bmatrix} = \begin{bmatrix} \quad & \quad \end{bmatrix}$$

Notice that the number of columns of A is 3 and it matches up with the number of rows of B. We need this because we're going to multiply each entry in the rows of A with the corresponding entry in the columns of B. For instance, consider the first row of A and the first column of B.

$$\begin{bmatrix} 1 & 3 & -1 \end{bmatrix} \begin{bmatrix} 1 \\ 1 \\ -1 \end{bmatrix}$$

The first entry in the first row of A is 1 and it corresponds to the first entry in the first column of B, which is 1. The second entry in the first row of A is 3 and it corresponds to the second entry in the first column of B, which is 1. The third entry in the first row of A is -1 and it corresponds to the third entry in the first column of B, which is -1.

To find the entry of AB in the first row and first column of AB, multiply out the corresponding entries and add: $1 \cdot 1 + 3 \cdot 1 + (-1) \cdot (-1) = 5$. So the product AB looks like this so far:

$$\begin{bmatrix} 5 & \quad \end{bmatrix}$$

To find the entry of AB in the first row and second column, use the first row of A and the second column of B, and perform the same procedure: $1 \cdot 0 + 3 \cdot 0 + (-1) \cdot 1 = -1$. So we get:

$$\begin{bmatrix} 5 & -1 \end{bmatrix}$$

Now, move on to the second row of A and the first column of B to get:

$$\begin{bmatrix} 5 & -1 \\ 0 & \quad \end{bmatrix}$$

Finally, move on to the second row of A and the second column of B:

$$\begin{bmatrix} 5 & -1 \\ 0 & 2 \end{bmatrix}$$

What if we had a matrix $C = \begin{bmatrix} 1 & 1 & -1 \\ 0 & 2 & 3 \end{bmatrix}$, which is a 2 × 3 matrix. Then we can't multiply AC:

$$\begin{bmatrix} 1 & 3 & -1 \\ 0 & 2 & 2 \end{bmatrix} \begin{bmatrix} 1 & 1 & -1 \\ 0 & 2 & 3 \end{bmatrix}$$

The number of columns of A is 3 and the number of rows of C is 2, and they don't match up. So we can't multiply them.

Now, for a general matrix, if the number of rows is the same as the number of columns, then the matrix is called a ***square matrix***. We can multiply two square matrices, of the same dimension, in any order. For example, suppose we had the following 2 × 2 square matrices:

$$A = \begin{bmatrix} 2 & 0 \\ 1 & -2 \end{bmatrix} \text{ and } B = \begin{bmatrix} 3 & -1 \\ 2 & 0 \end{bmatrix}$$

We can find AB:

$$\begin{bmatrix} 2 & 0 \\ 1 & -2 \end{bmatrix} \begin{bmatrix} 3 & -1 \\ 2 & 0 \end{bmatrix} = \begin{bmatrix} 6 & -2 \\ -1 & -1 \end{bmatrix}$$

We can also multiply them in reverse order:

$$BA = \begin{bmatrix} 3 & -1 \\ 2 & 0 \end{bmatrix} \begin{bmatrix} 2 & 0 \\ 1 & -2 \end{bmatrix} = \begin{bmatrix} 5 & 2 \\ 4 & 0 \end{bmatrix}$$

Note that, in this case, $AB \neq BA$. In general, matrix multiplication is not commutative.

PROBLEM SET: MATRIX OPERATIONS

Calculate A+B and A-B.

1. $A = \begin{bmatrix} 3 & 4 \\ 5 & 1 \end{bmatrix} \qquad B = \begin{bmatrix} -1 & 0 \\ 1 & 1 \end{bmatrix}$

2. $A = \begin{bmatrix} 8 & 0 \\ -1 & 0 \end{bmatrix} \qquad B = \begin{bmatrix} 2 & 2 \\ 10 & -1 \end{bmatrix}$

Find cA.

1. $c=2 \quad A = \begin{bmatrix} -1 & 0 \\ -1 & 1 \end{bmatrix}$

2. $c=8 \quad A = \begin{bmatrix} 4 & 3 & 1 \\ 0 & 0 & 2 \\ -3 & -3 & 9 \end{bmatrix}$

Find AB if A and B can be multiplied. Otherwise, indicate why they cannot be multiplied.

1. $A = \begin{bmatrix} 2 & 0 \\ -1 & 1 \end{bmatrix} \qquad B = \begin{bmatrix} 3 & 3 \\ 4 & 1 \end{bmatrix}$

2. $A = \begin{bmatrix} 3 & 0 & 1 \\ -1 & 1 & 1 \\ 8 & 7 & 6 \end{bmatrix} \qquad B = \begin{bmatrix} 2 & -2 & 3 \\ 6 & 6 & 2 \\ 1 & 4 & 4 \end{bmatrix}$

3. $A = \begin{bmatrix} 0 & 1 \\ 1 & 1 \end{bmatrix} \qquad B = \begin{bmatrix} 3 & 1 & -1 \\ 1 & 1 & 2 \\ 0 & 1 & 2 \end{bmatrix}$

4. $A = \begin{bmatrix} 2 & 0 & 1 \\ 1 & -1 & 1 \end{bmatrix} \qquad B = \begin{bmatrix} 3 & -1 & 1 \\ 1 & 1 & 4 \\ 0 & 1 & 3 \end{bmatrix}$

5. $A = \begin{bmatrix} 1 & 1 \\ 2 & -1 \\ 1 & 0 \end{bmatrix} \qquad B = \begin{bmatrix} 0 & 0 & 1 \\ 7 & 6 & 5 \end{bmatrix}$

SOLUTION SET: MATRIX OPERATIONS

Calculate A+B and A-B.

3. $A + B = \begin{bmatrix} 2 & 4 \\ 6 & 2 \end{bmatrix}$ $A - B = \begin{bmatrix} 4 & 4 \\ 4 & 0 \end{bmatrix}$

4. $A + B = \begin{bmatrix} 10 & 2 \\ 9 & -1 \end{bmatrix}$ $A - B = \begin{bmatrix} 6 & -2 \\ -11 & 1 \end{bmatrix}$

Find cA.

3. $cA = \begin{bmatrix} -2 & 0 \\ -2 & 2 \end{bmatrix}$

4. $cA = \begin{bmatrix} 32 & 24 & 8 \\ 0 & 0 & 16 \\ -24 & -24 & 72 \end{bmatrix}$

Find AB if A and B can be multiplied. Otherwise, indicate why they cannot be multiplied.

6. $AB = \begin{bmatrix} 6 & 6 \\ 1 & -2 \end{bmatrix}$

7. $AB = \begin{bmatrix} 7 & -2 & 13 \\ 5 & 12 & 3 \\ 64 & 50 & 62 \end{bmatrix}$

8. A is 2 by 2 and B is 3 by 3.

9. $AB = \begin{bmatrix} 6 & -1 & 5 \\ 2 & -1 & 0 \end{bmatrix}$

10. $AB = \begin{bmatrix} 7 & 6 & 6 \\ -7 & -6 & -3 \\ 0 & 0 & 1 \end{bmatrix}$

SUMMARY: MATRIX OPERATIONS

- When adding two matrices, we simply add the corresponding entries. When multiplying a matrix by a scalar, we multiply each entry of the matrix by the scalar.

- To multiply two matrices A and B, the number of columns of A has to be the same as the number of rows of B. To find the ij-th entry of AB, we take the entries of the i-th row of A and multiply by the corresponding entries of the j-th column of B, then add the results.

- A square matrix is a matrix with the same number of rows and columns.

5 – PROPERTIES OF MATRIX ADDITION AND SCALAR MULTIPLICATION

COMMUTATIVITY, ASSOCIATIVITY, AND DISTRIBUTIVITY

In this section, we're going to look at properties of matrix addition and scalar multiplication. The first two properties we're going to look at are commutativity and associativity.

Matrix addition is commutative and associative. The commutativity property states that $A + B = B + A$. So, the order does not matter when adding. The associativity property states that $A + (B + C) = (A + B) + C$. So, we can add B and C first before adding A or we can add A and B first before adding C.

Let's do some examples. Suppose $= \begin{bmatrix} 1 & 0 \\ 0 & -1 \end{bmatrix}, B = \begin{bmatrix} 2 & 1 \\ 3 & 4 \end{bmatrix}, C = \begin{bmatrix} -1 & 7 \\ 13 & 5 \end{bmatrix}$.

Then, $A + B = \begin{bmatrix} 1 & 0 \\ 0 & -1 \end{bmatrix} + \begin{bmatrix} 2 & 1 \\ 3 & 4 \end{bmatrix} = \begin{bmatrix} 3 & 1 \\ 3 & 3 \end{bmatrix}$.

Now find $B + A = \begin{bmatrix} 2 & 1 \\ 3 & 4 \end{bmatrix} + \begin{bmatrix} 1 & 0 \\ 0 & -1 \end{bmatrix} = \begin{bmatrix} 3 & 1 \\ 3 & 3 \end{bmatrix}$. Note that we get the same result. So this illustrates commutativity of addition.

Let's do an example of associativity. Let's find $A + (B + C)$.

$$A + (B + C) = \begin{bmatrix} 1 & 0 \\ 0 & -1 \end{bmatrix} + \left(\begin{bmatrix} 2 & 1 \\ 3 & 4 \end{bmatrix} + \begin{bmatrix} -1 & 7 \\ 13 & 5 \end{bmatrix} \right)$$

$$= \begin{bmatrix} 1 & 0 \\ 0 & -1 \end{bmatrix} + \begin{bmatrix} 1 & 8 \\ 16 & 9 \end{bmatrix} \quad \text{(adding the matrices inside the parentheses)}$$

$$= \begin{bmatrix} 2 & 8 \\ 16 & 8 \end{bmatrix}$$

Now let's do $(A + B) + C$ and see if we get the same thing.

$$(A + B) + C = \left(\begin{bmatrix} 1 & 0 \\ 0 & -1 \end{bmatrix} + \begin{bmatrix} 2 & 1 \\ 3 & 4 \end{bmatrix} \right) + \begin{bmatrix} -1 & 7 \\ 13 & 5 \end{bmatrix}$$

$$= \begin{bmatrix} 3 & 1 \\ 3 & 3 \end{bmatrix} + \begin{bmatrix} -1 & 7 \\ 13 & 5 \end{bmatrix}$$

$$= \begin{bmatrix} 2 & 8 \\ 16 & 8 \end{bmatrix}$$

Note that we get the same result as $A + (B + C)$. This illustrates associativity of addition.

Let's look at another property.

Let c and d be scalars. Then $(cd)A = c(dA)$.

For example, let $c = 2, d = -1, A = \begin{bmatrix} 1 & 0 \\ 0 & -1 \end{bmatrix}$. Then $cd = -2$ and

$$(cd)A = -2\begin{bmatrix} 1 & 0 \\ 0 & -1 \end{bmatrix} = \begin{bmatrix} -2 & 0 \\ 0 & 2 \end{bmatrix}.$$

Now let's find $c(dA)$.

$$c(dA) = 2\left(-1\begin{bmatrix} 1 & 0 \\ 0 & -1 \end{bmatrix}\right) = 2\begin{bmatrix} -1 & 0 \\ 0 & 1 \end{bmatrix} = \begin{bmatrix} -2 & 0 \\ 0 & 2 \end{bmatrix}.$$

Note that we get the same result as $(cd)A$.

Next, we have the distributivity properties:

$c(A + B) = cA + cB$

$(c + d)A = cA + dA$

Let's do an example. Let $c = 2, A = \begin{bmatrix} 1 & 0 \\ 0 & -1 \end{bmatrix}, B = \begin{bmatrix} 2 & 1 \\ 3 & 4 \end{bmatrix}$.

Then, $c(A + B) = 2\left(\begin{bmatrix} 1 & 0 \\ 0 & -1 \end{bmatrix} + \begin{bmatrix} 2 & 1 \\ 3 & 4 \end{bmatrix}\right)$

$$= 2\begin{bmatrix} 3 & 1 \\ 3 & 3 \end{bmatrix}$$

$$= \begin{bmatrix} 6 & 2 \\ 6 & 6 \end{bmatrix}$$

Now, $cA + cB = 2\begin{bmatrix} 1 & 0 \\ 0 & -1 \end{bmatrix} + 2\begin{bmatrix} 2 & 1 \\ 3 & 4 \end{bmatrix}$

$$= \begin{bmatrix} 2 & 0 \\ 0 & -2 \end{bmatrix} + \begin{bmatrix} 4 & 2 \\ 6 & 8 \end{bmatrix}$$

$$= \begin{bmatrix} 6 & 2 \\ 6 & 6 \end{bmatrix}$$

So, we get the same result as $c(A + B)$.

IDENTITIES, ADDITIVE INVERSES, MULTIPLICATIVE ASSOCIATIVITY AND DISTRIBUTIVITY

There's an additive identity for matrices. The **additive identity** is the matrix with zeroes everywhere. So, for a 2 × 2 matrix, it looks like this: $\begin{bmatrix} 0 & 0 \\ 0 & 0 \end{bmatrix}$.

The matrix with all zeroes is called the **zero matrix**. If we add the zero matrix to any matrix, then we get back the same matrix. It's like adding 0 to a number.

Here's an example: $\begin{bmatrix} 0 & 0 \\ 0 & 0 \end{bmatrix} + \begin{bmatrix} 1 & 0 \\ 0 & -1 \end{bmatrix} = \begin{bmatrix} 1 & 0 \\ 0 & -1 \end{bmatrix}$.

Additive inverses exist for matrices, and the **additive inverse** of A is $-A$.

For example, $\begin{bmatrix} 1 & 0 \\ 0 & -1 \end{bmatrix} + -\begin{bmatrix} 1 & 0 \\ 0 & -1 \end{bmatrix} = \begin{bmatrix} 1 & 0 \\ 0 & -1 \end{bmatrix} + \begin{bmatrix} -1 & 0 \\ 0 & 1 \end{bmatrix} = \begin{bmatrix} 0 & 0 \\ 0 & 0 \end{bmatrix}$.

Adding the additive inverse of A to A gives us the zero matrix.

Associativity and distributivity hold for matrix multiplication.

The associativity property for multiplication states $A(BC) = (AB)C$.

Distributivity states that $A(B + C) = AB + AC$. The A distributes to B and C.

We also have distributivity from the right: $(A + B)C = AC + BC$. The C distributes to A and B from the right.

We also have this property, where c is a scalar: $c(AB) = (cA)B = A(cB)$. We can pull the scalar c out.

Let's do some examples. Here's an example of associativity.

Suppose $A = \begin{bmatrix} 1 & 0 \\ 0 & -1 \end{bmatrix}, B = \begin{bmatrix} 2 & 1 \\ 3 & 4 \end{bmatrix}, C = \begin{bmatrix} -1 & 7 \\ 13 & 5 \end{bmatrix}$. Let's show that $(AB)C = A(BC)$.

$$(AB)C = \left(\begin{bmatrix} 1 & 0 \\ 0 & -1 \end{bmatrix}\begin{bmatrix} 2 & 1 \\ 3 & 4 \end{bmatrix}\right)\begin{bmatrix} -1 & 7 \\ 13 & 5 \end{bmatrix}$$

$$= \begin{bmatrix} 2 & 1 \\ -3 & -4 \end{bmatrix}\begin{bmatrix} -1 & 7 \\ 13 & 5 \end{bmatrix}$$

$$= \begin{bmatrix} 11 & 19 \\ -49 & -41 \end{bmatrix}$$

$$A(BC) = \begin{bmatrix} 1 & 0 \\ 0 & -1 \end{bmatrix}\left(\begin{bmatrix} 2 & 1 \\ 3 & 4 \end{bmatrix}\begin{bmatrix} -1 & 7 \\ 13 & 5 \end{bmatrix}\right)$$

$$= \begin{bmatrix} 1 & 0 \\ 0 & -1 \end{bmatrix}\begin{bmatrix} 11 & 19 \\ 49 & 41 \end{bmatrix}$$

$$= \begin{bmatrix} 11 & 19 \\ -49 & -41 \end{bmatrix}$$

Let's do an example showing $c(AB) = (cA)B = A(cB)$.

Suppose $c = 3, A = \begin{bmatrix} 1 & 0 \\ 0 & -1 \end{bmatrix}, B = \begin{bmatrix} 2 & 1 \\ 3 & 4 \end{bmatrix}$.

$$c(AB) = 3\left(\begin{bmatrix} 1 & 0 \\ 0 & -1 \end{bmatrix}\begin{bmatrix} 2 & 1 \\ 3 & 4 \end{bmatrix}\right)$$

$$= 3\begin{bmatrix} 2 & 1 \\ -3 & -4 \end{bmatrix}$$

$$= \begin{bmatrix} 6 & 3 \\ -9 & -12 \end{bmatrix}$$

Now, calculate $(cA)B$:

$$(cA)B = \left(3\begin{bmatrix} 1 & 0 \\ 0 & -1 \end{bmatrix}\right)\begin{bmatrix} 2 & 1 \\ 3 & 4 \end{bmatrix}$$

$$= \begin{bmatrix} 3 & 0 \\ 0 & -3 \end{bmatrix}\begin{bmatrix} 2 & 1 \\ 3 & 4 \end{bmatrix}$$

$$= \begin{bmatrix} 6 & 3 \\ -9 & -12 \end{bmatrix}$$

Now, calculate $A(cB)$:

$$A(cB) = \begin{bmatrix} 1 & 0 \\ 0 & -1 \end{bmatrix}\left(3\begin{bmatrix} 2 & 1 \\ 3 & 4 \end{bmatrix}\right)$$

$$= \begin{bmatrix} 1 & 0 \\ 0 & -1 \end{bmatrix}\begin{bmatrix} 6 & 3 \\ 9 & 12 \end{bmatrix}$$

$$= \begin{bmatrix} 6 & 3 \\ -9 & -12 \end{bmatrix}$$

We get the same result in all three cases.

There is a **multiplicative identity** element for matrices, and it is denoted by I. It has 1's along the diagonal and zeroes everywhere else. For instance, for 2×2 matrices, $I = \begin{bmatrix} 1 & 0 \\ 0 & 1 \end{bmatrix}$. Just as multiplying a number by 1 gives back the same number, multiplying a matrix by I gives back the same matrix. For instance, $IA = \begin{bmatrix} 1 & 0 \\ 0 & 1 \end{bmatrix}\begin{bmatrix} 1 & 0 \\ 0 & -1 \end{bmatrix} = \begin{bmatrix} 1 & 0 \\ 0 & -1 \end{bmatrix}$, and $AI = \begin{bmatrix} 1 & 0 \\ 0 & -1 \end{bmatrix}\begin{bmatrix} 1 & 0 \\ 0 & 1 \end{bmatrix} = \begin{bmatrix} 1 & 0 \\ 0 & -1 \end{bmatrix}$.

PROBLEM SET: PROPERTIES OF MATRIX OPERATIONS

1. Show that commutativity under addition holds for A and B.

$$A = \begin{bmatrix} 1 & 7 & 8 \\ -7 & 6 & 10 \end{bmatrix} \qquad B = \begin{bmatrix} 0 & 1 & 2 \\ 2 & -2 & 1 \end{bmatrix}$$

2. Show that multiplicative associativity holds for A, B, and C.

$$A = \begin{bmatrix} -1 & 0 \\ 1 & 2 \end{bmatrix} \qquad B = \begin{bmatrix} 2 & 2 \\ 2 & 2 \end{bmatrix} \qquad C = \begin{bmatrix} 0 & 1 \\ 1 & 0 \end{bmatrix}$$

3. Show that $AB \neq BA$ for the following matrices:

$$A = \begin{bmatrix} 8 & -8 \\ 1 & 1 \end{bmatrix} \qquad B = \begin{bmatrix} 2 & 1 \\ 3 & 4 \end{bmatrix}$$

4. Show that (c+d)A=cA+dA.

$$c=4, \quad d=2, \quad A = \begin{bmatrix} 1 & 0 & -3 \\ 3 & 2 & 1 \\ -1 & -1 & 0 \end{bmatrix}$$

5. Show that (A+B)C=AC+BC

$$A = \begin{bmatrix} 1 & 2 & 3 \\ 4 & 0 & 1 \end{bmatrix} \qquad B = \begin{bmatrix} -1 & 0 & 3 \\ 4 & 1 & 4 \end{bmatrix} \qquad C = \begin{bmatrix} 1 & 0 \\ 0 & 1 \\ -1 & 0 \end{bmatrix}$$

6. Show that AI=A, where I is the identity matrix.

$$A = \begin{bmatrix} 1 & 8 & 8 \\ 13 & 24 & 6 \\ 6 & -6 & 8 \end{bmatrix}$$

SOLUTION SET: PROPERTIES OF MATRIX OPERATIONS

1. Show that commutativity under addition holds for A and B.

$$A = \begin{bmatrix} 1 & 7 & 8 \\ -7 & 6 & 10 \end{bmatrix} \quad B = \begin{bmatrix} 0 & 1 & 2 \\ 2 & -2 & 1 \end{bmatrix}$$

$$A + B = \begin{bmatrix} 1 & 7 & 8 \\ -7 & 6 & 10 \end{bmatrix} + \begin{bmatrix} 0 & 1 & 2 \\ 2 & -2 & 1 \end{bmatrix} = \begin{bmatrix} 1+0 & 7+1 & 8+2 \\ -7+2 & 6-2 & 10+1 \end{bmatrix} = \begin{bmatrix} 1 & 8 & 10 \\ -5 & 4 & 11 \end{bmatrix}$$

$$B + A = \begin{bmatrix} 0 & 1 & 2 \\ 2 & -2 & 1 \end{bmatrix} + \begin{bmatrix} 1 & 7 & 8 \\ -7 & 6 & 10 \end{bmatrix} = \begin{bmatrix} 0+1 & 1+7 & 2+8 \\ 2-7 & -2+6 & 1+10 \end{bmatrix} = \begin{bmatrix} 1 & 8 & 10 \\ -5 & 4 & 11 \end{bmatrix}$$

So $A + B = B + A$.

2. Show that multiplicative associativity holds for A, B, and C.

$$A = \begin{bmatrix} -1 & 0 \\ 1 & 2 \end{bmatrix} \quad B = \begin{bmatrix} 2 & 2 \\ 2 & 2 \end{bmatrix} \quad C = \begin{bmatrix} 0 & 1 \\ 1 & 0 \end{bmatrix}$$

$$(AB)C = \left(\begin{bmatrix} -1 & 0 \\ 1 & 2 \end{bmatrix}\begin{bmatrix} 2 & 2 \\ 2 & 2 \end{bmatrix}\right)\begin{bmatrix} 0 & 1 \\ 1 & 0 \end{bmatrix} = \begin{bmatrix} -2 & -2 \\ 6 & 6 \end{bmatrix}\begin{bmatrix} 0 & 1 \\ 1 & 0 \end{bmatrix} = \begin{bmatrix} -2 & -2 \\ 6 & 6 \end{bmatrix}$$

$$A(BC) = \begin{bmatrix} -1 & 0 \\ 1 & 2 \end{bmatrix}\left(\begin{bmatrix} 2 & 2 \\ 2 & 2 \end{bmatrix}\begin{bmatrix} 0 & 1 \\ 1 & 0 \end{bmatrix}\right) = \begin{bmatrix} -1 & 0 \\ 1 & 2 \end{bmatrix}\begin{bmatrix} 2 & 2 \\ 2 & 2 \end{bmatrix} = \begin{bmatrix} -2 & -2 \\ 6 & 6 \end{bmatrix}$$

So $(AB)C = A(BC)$.

3. Show that $AB \neq BA$ for the following matrices:

$$A = \begin{bmatrix} 8 & -8 \\ 1 & 1 \end{bmatrix} \quad B = \begin{bmatrix} 2 & 1 \\ 3 & 4 \end{bmatrix}$$

$$AB = \begin{bmatrix} 8 & -8 \\ 1 & 1 \end{bmatrix}\begin{bmatrix} 2 & 1 \\ 3 & 4 \end{bmatrix} = \begin{bmatrix} -8 & -24 \\ 5 & 5 \end{bmatrix}$$

$$BA = \begin{bmatrix} 2 & 1 \\ 3 & 4 \end{bmatrix}\begin{bmatrix} 8 & -8 \\ 1 & 1 \end{bmatrix} = \begin{bmatrix} 17 & -15 \\ 28 & -20 \end{bmatrix}$$

So $AB \neq BA$.

4. Show that (c+d)A=cA+dA.

$$c=4, \quad d=2, \quad A = \begin{bmatrix} 1 & 0 & -3 \\ 3 & 2 & 1 \\ -1 & -1 & 0 \end{bmatrix}$$

$$(c+d)A = (4+2)A = 6A = 6\begin{bmatrix} 1 & 0 & -3 \\ 3 & 2 & 1 \\ -1 & -1 & 0 \end{bmatrix}$$

$$= \begin{bmatrix} 6 & 0 & -18 \\ 18 & 12 & 6 \\ -6 & -6 & 0 \end{bmatrix}$$

$$cA + dA = 4A + 2A = 4\begin{bmatrix} 1 & 0 & -3 \\ 3 & 2 & 1 \\ -1 & -1 & 0 \end{bmatrix} + 2\begin{bmatrix} 1 & 0 & -3 \\ 3 & 2 & 1 \\ -1 & -1 & 0 \end{bmatrix}$$

$$= \begin{bmatrix} 4 & 0 & -12 \\ 12 & 8 & 4 \\ -4 & -4 & 0 \end{bmatrix} + \begin{bmatrix} 2 & 0 & -6 \\ 6 & 4 & 2 \\ -2 & -2 & 0 \end{bmatrix} = \begin{bmatrix} 6 & 0 & -18 \\ 18 & 12 & 6 \\ -6 & -6 & 0 \end{bmatrix}$$

So $(c + d)A = cA + dA$.

5. Show that (A+B)C=AC+BC

$$A = \begin{bmatrix} 1 & 2 & 3 \\ 4 & 0 & 1 \end{bmatrix} \quad B = \begin{bmatrix} -1 & 0 & 3 \\ 4 & 1 & 4 \end{bmatrix} \quad C = \begin{bmatrix} 1 & 0 \\ 0 & 1 \\ -1 & 0 \end{bmatrix}$$

$$(A + B)C = \left(\begin{bmatrix} 1 & 2 & 3 \\ 4 & 0 & 1 \end{bmatrix} + \begin{bmatrix} -1 & 0 & 3 \\ 4 & 1 & 4 \end{bmatrix}\right)\begin{bmatrix} 1 & 0 \\ 0 & 1 \\ -1 & 0 \end{bmatrix}$$

$$= \begin{bmatrix} 0 & 2 & 6 \\ 8 & 1 & 5 \end{bmatrix}\begin{bmatrix} 1 & 0 \\ 0 & 1 \\ -1 & 0 \end{bmatrix}$$

$$= \begin{bmatrix} -6 & 2 \\ 3 & 1 \end{bmatrix}$$

$$AC + BC = \begin{bmatrix} 1 & 2 & 3 \\ 4 & 0 & 1 \end{bmatrix}\begin{bmatrix} 1 & 0 \\ 0 & 1 \\ -1 & 0 \end{bmatrix} + \begin{bmatrix} -1 & 0 & 3 \\ 4 & 1 & 4 \end{bmatrix}\begin{bmatrix} 1 & 0 \\ 0 & 1 \\ -1 & 0 \end{bmatrix}$$

$$= \begin{bmatrix} -2 & 2 \\ 3 & 0 \end{bmatrix} + \begin{bmatrix} -4 & 0 \\ 0 & 1 \end{bmatrix}$$

$$= \begin{bmatrix} -6 & 2 \\ 3 & 1 \end{bmatrix}$$

So $(A + B)C = AC + BC$.

6. Show that AI=A, where I is the identity matrix.

$$A = \begin{bmatrix} 1 & 8 & 8 \\ 13 & 24 & 6 \\ 6 & -6 & 8 \end{bmatrix}$$

$$AI = \begin{bmatrix} 1 & 8 & 8 \\ 13 & 24 & 6 \\ 6 & -6 & 8 \end{bmatrix}\begin{bmatrix} 1 & 0 & 0 \\ 0 & 1 & 0 \\ 0 & 0 & 1 \end{bmatrix}$$

$$= \begin{bmatrix} 1\cdot 1+8\cdot 0+8\cdot 0 & 1\cdot 0+8\cdot 1+8\cdot 0 & 1\cdot 0+8\cdot 0+8\cdot 1 \\ 13\cdot 1+24\cdot 0+6\cdot 0 & 13\cdot 0+24\cdot 1+6\cdot 0 & 13\cdot 0+24\cdot 0+6\cdot 1 \\ 6\cdot 1-6\cdot 0+8\cdot 0 & 6\cdot 0-6\cdot 1+8\cdot 0 & 6\cdot 0-6\cdot 0+8\cdot 1 \end{bmatrix}$$

$$= \begin{bmatrix} 1 & 8 & 8 \\ 13 & 24 & 6 \\ 6 & -6 & 8 \end{bmatrix}$$

$= A.$

TRANSPOSE OF A MATRIX

We're now going to learn about the transpose of a matrix. The *transpose* of a matrix is the matrix you get by swapping the columns and rows of the matrix.

For example, suppose $A = \begin{bmatrix} 1 & 0 \\ -13 & 8 \end{bmatrix}$. Then, $A^T = \begin{bmatrix} 1 & -13 \\ 0 & 8 \end{bmatrix}$. The first row of A becomes the first column of A^T, and the second row of A becomes the second column of A^T.

Let's look at another example. Suppose $B = \begin{bmatrix} 2 & -1 & 3 \\ 4 & 1 & 1 \\ 8 & 9 & 3 \end{bmatrix}$. Then, $B^T = \begin{bmatrix} 2 & 4 & 8 \\ -1 & 1 & 9 \\ 3 & 1 & 3 \end{bmatrix}$.

Now, the transpose satisfies the following properties:

$(A^T)^T = A$ $\qquad\qquad (cA)^T = c(A^T)$

$(A + B)^T = A^T + B^T \qquad (AB)^T = B^T A^T$

If you take the transpose of the transpose of A, you get back A. If you take the sum $A + B$ and take the transpose of that, it's the same as taking the transposes individually and adding. If you multiply A by a scalar c and take the transpose of the result, it's the same as first taking the transpose of A and then multiplying by c. Taking the transpose of a product is the same as taking the product of the transposes but in reverse order.

Let's see an example. Let $A = \begin{bmatrix} 1 & 0 \\ -13 & 8 \end{bmatrix}$ and $B = \begin{bmatrix} 2 & 1 \\ 1 & 2 \end{bmatrix}$.

$$AB = \begin{bmatrix} 1 & 0 \\ -13 & 8 \end{bmatrix}\begin{bmatrix} 2 & 1 \\ 1 & 2 \end{bmatrix} = \begin{bmatrix} 2 & 1 \\ -18 & 3 \end{bmatrix}$$

Taking the transpose, we get $(AB)^T = \begin{bmatrix} 2 & -18 \\ 1 & 3 \end{bmatrix}$. Let's check that this is equal to $B^T A^T$.

$$B^T A^T = \begin{bmatrix} 2 & 1 \\ 1 & 2 \end{bmatrix}\begin{bmatrix} 1 & -13 \\ 0 & 8 \end{bmatrix} = \begin{bmatrix} 2 & -18 \\ 1 & 3 \end{bmatrix}$$

PROBLEM SET: TRANSPOSE OF A MATRIX

1. Show that $(A+B)^T = A^T + B^T$

   $A = \begin{bmatrix} 1 & 0 & -3 \\ 3 & 6 & 8 \end{bmatrix}$ $B = \begin{bmatrix} 1 & -1 & 1 \\ 0 & 0 & 1 \end{bmatrix}.$

2. Show that $(cA)^T = c(A^T)$

   $c=4, A = \begin{bmatrix} 1 & 2 \\ 0 & 2 \\ -1 & 2 \end{bmatrix}$

3. Show that $(A^T)^T = A.$

   $A = \begin{bmatrix} 0 & 1 & 6 \\ 4 & 5 & 6 \end{bmatrix}$

SOLUTION SET: TRANSPOSE OF A MATRIX

1. Show that $(A+B)^T = A^T + B^T$

    $A = \begin{bmatrix} 1 & 0 & -3 \\ 3 & 6 & 8 \end{bmatrix}$ $\qquad B = \begin{bmatrix} 1 & -1 & 1 \\ 0 & 0 & 1 \end{bmatrix}$.

    $(A+B) = \begin{bmatrix} 2 & -1 & -2 \\ 3 & 6 & 9 \end{bmatrix}$. To find the transpose of (A+B), switch the rows and columns.

    $(A+B)^T = \begin{bmatrix} 2 & 3 \\ -1 & 6 \\ -2 & 9 \end{bmatrix}$.

    $A^T = \begin{bmatrix} 1 & 3 \\ 0 & 6 \\ -3 & 8 \end{bmatrix}$ and $B^T = \begin{bmatrix} 1 & 0 \\ -1 & 0 \\ 1 & 1 \end{bmatrix}$. If we add A^T and B^T, we get $\begin{bmatrix} 2 & 3 \\ -1 & 6 \\ -2 & 9 \end{bmatrix}$. This is the same as $(A+B)^T$.

2. Show that $(cA)^T = c(A^T)$

    $c = 4,\ A = \begin{bmatrix} 1 & 2 \\ 0 & 2 \\ -1 & 2 \end{bmatrix}$

    $cA = 4 \begin{bmatrix} 1 & 2 \\ 0 & 2 \\ -1 & 2 \end{bmatrix} = \begin{bmatrix} 4 & 8 \\ 0 & 8 \\ -4 & 8 \end{bmatrix}$. The transpose of this is $(cA)^T = \begin{bmatrix} 4 & 0 & -4 \\ 8 & 8 & 8 \end{bmatrix}$.

    The transpose of A is $\begin{bmatrix} 1 & 0 & -1 \\ 2 & 2 & 2 \end{bmatrix}$. Now, $c(A^T) = 4 \begin{bmatrix} 1 & 0 & -1 \\ 2 & 2 & 2 \end{bmatrix} = \begin{bmatrix} 4 & 0 & -4 \\ 8 & 8 & 8 \end{bmatrix}$. This is the same as $(cA)^T$.

3. Show that $(A^T)^T = A$.

    $A = \begin{bmatrix} 0 & 1 & 6 \\ 4 & 5 & 6 \end{bmatrix}$

    $A^T = \begin{bmatrix} 0 & 4 \\ 1 & 5 \\ 6 & 6 \end{bmatrix}$. The transpose of this is $(A^T)^T = \begin{bmatrix} 0 & 1 & 6 \\ 4 & 5 & 6 \end{bmatrix}$. This is the same as A.

SUMMARY: PROPERTIES OF MATRIX ADDITION AND SCALAR MULTIPLICATION

- The commutativity property of addition states $A + B = B + A$.

- The associativity property of addition states $A + (B + C) = (A + B) + C$.

- If c and d are scalars, $(cd)A = c(dA)$.

- The distributivity properties state:
$$c(A + B) = cA + cB$$
$$(c + d)A = cA + dA$$

- The *additive identity* is the matrix with zeroes everywhere.

- The matrix with all zeroes is called the *zero matrix.*

- Additive inverses exist for matrices, and the *additive inverse* of A is $-A$.

- The associativity property of multiplication states $A(BC) = (AB)C$.

- The distributivity properties for multiplication state:
$$A(B + C) = AB + AC$$
$$(A + B)C = AC + BC$$

- If c is a scalar, $c(AB) = (cA)B = A(cB)$.

- There is a *multiplicative identity* element for matrices, and it is denoted by I. It has 1's along the diagonal and zeroes everywhere else.

- The *transpose* of a matrix is the matrix you get by swapping the columns and rows of the matrix.

- The transpose satisfies the following properties:
$(A^T)^T = A$ $\qquad$ $(cA)^T = c(A^T)$
$(A + B)^T = A^T + B^T$ $\qquad$ $(AB)^T = B^T A^T$

6 – THE INVERSE OF A MATRIX

INVERSE MATRIX

In this section, we're going to look at the inverse of a matrix. We've explored addition, scalar multiplication, subtraction, and matrix multiplication for matrices. We've seen that a matrix always has an additive inverse. We might wonder if a matrix has a multiplicative inverse. The *inverse* of a matrix A is any matrix B such that when you multiply A on the left and right by the matrix B, you get the identity matrix. So it looks like this:

$$AB = BA = I$$

The inverse is denoted A^{-1}. In matrix algebra, there is no division, but the analogue is the inverse matrix. Let's do an example.

Suppose $A = \begin{bmatrix} 4 & 0 \\ -1 & 2 \end{bmatrix}$. For 2×2 matrices, there is a formula for the inverse. If $A = \begin{bmatrix} a & b \\ c & d \end{bmatrix}$, then $A^{-1} = \frac{1}{ad-bc}\begin{bmatrix} d & -b \\ -c & a \end{bmatrix}$. Applying the inverse formula to our matrix A, we get:

$$\frac{1}{8}\begin{bmatrix} 2 & 0 \\ 1 & 4 \end{bmatrix}$$

$$= \begin{bmatrix} 1/4 & 0 \\ 1/8 & 1/2 \end{bmatrix}$$

We can check that this is the inverse of A by checking that $AA^{-1} = I$ and $A^{-1}A = I$.

$$AA^{-1} = \begin{bmatrix} 4 & 0 \\ -1 & 2 \end{bmatrix}\begin{bmatrix} 1/4 & 0 \\ 1/8 & 1/2 \end{bmatrix} = \begin{bmatrix} 1 & 0 \\ 0 & 1 \end{bmatrix}$$

$$A^{-1}A = \begin{bmatrix} 1/4 & 0 \\ 1/8 & 1/2 \end{bmatrix}\begin{bmatrix} 4 & 0 \\ -1 & 2 \end{bmatrix} = \begin{bmatrix} 1 & 0 \\ 0 & 1 \end{bmatrix}$$

GAUSS-JORDAN ELIMINATION

To find the inverse of a matrix, we can use a process called **Gauss-Jordan elimination**. We take the matrix A and adjoin the identity matrix. Then perform row operations on the resulting matrix until we transform A into I. Let's do an example.

Suppose $A = \begin{bmatrix} 4 & 0 \\ -1 & 2 \end{bmatrix}$. We take A and adjoin the identity matrix like this:

$$\begin{bmatrix} 4 & 0 & : & 1 & 0 \\ -1 & 2 & : & 0 & 1 \end{bmatrix}$$

Then we start performing row operations on this matrix until we get the left hand side to look like the right hand side. Let's do $\frac{1}{4}R1 \rightarrow R1$:

$$\begin{bmatrix} 1 & 0 & : & 1/4 & 0 \\ -1 & 2 & : & 0 & 1 \end{bmatrix}$$

Now do $R1 + R2 \rightarrow R2$:

$$\begin{bmatrix} 1 & 0 & : & 1/4 & 0 \\ 0 & 2 & : & 1/4 & 1 \end{bmatrix}$$

Now do $\frac{1}{2}R2 \rightarrow R2$:

$$\begin{bmatrix} 1 & 0 & : & 1/4 & 0 \\ 0 & 1 & : & 1/8 & 1/2 \end{bmatrix}$$

Now that we have the identity matrix on the left hand side, we look at the matrix on the right hand side. That will be our inverse matrix. So $A^{-1} = \begin{bmatrix} 1/4 & 0 \\ 1/8 & 1/2 \end{bmatrix}$.

Let's do another example.

Suppose $A = \begin{bmatrix} 1 & 0 & -1 \\ 2 & 2 & 1 \\ 0 & 0 & 3 \end{bmatrix}$.

Form the matrix with A on the left and the identity on the right like this:

$$\begin{bmatrix} 1 & 0 & -1 & : & 1 & 0 & 0 \\ 2 & 2 & 1 & : & 0 & 1 & 0 \\ 0 & 0 & 3 & : & 0 & 0 & 1 \end{bmatrix}$$

Then start performing row operations until we transform the left hand side into the identity matrix.

Let's do $-2R1 + R2 \rightarrow R2$:

$$\begin{bmatrix} 1 & 0 & -1 & : & 1 & 0 & 0 \\ 0 & 2 & 3 & : & -2 & 1 & 0 \\ 0 & 0 & 3 & : & 0 & 0 & 1 \end{bmatrix}$$

Let's do $\frac{1}{3}R3 \rightarrow R3$:

$$\begin{bmatrix} 1 & 0 & -1 & : & 1 & 0 & 0 \\ 0 & 2 & 3 & : & -2 & 1 & 0 \\ 0 & 0 & 1 & : & 0 & 0 & 1/3 \end{bmatrix}$$

Similarly, let's do $\frac{1}{2}R2 \rightarrow R2$:

$$\begin{bmatrix} 1 & 0 & -1 & : & 1 & 0 & 0 \\ 0 & 1 & 3/2 & : & -1 & 1/2 & 0 \\ 0 & 0 & 1 & : & 0 & 0 & 1/3 \end{bmatrix}$$

Let's do $-\frac{3}{2}R3 + R2 \rightarrow R2$:

$$\begin{bmatrix} 1 & 0 & -1 & : & 1 & 0 & 0 \\ 0 & 1 & 0 & : & -1 & 1/2 & -1/2 \\ 0 & 0 & 1 & : & 0 & 0 & 1/3 \end{bmatrix}$$

Let's do $R1 + R3 \rightarrow R1$:

$$\begin{bmatrix} 1 & 0 & 0 & : & 1 & 0 & 1/3 \\ 0 & 1 & 0 & : & -1 & 1/2 & -1/2 \\ 0 & 0 & 1 & : & 0 & 0 & 1/3 \end{bmatrix}$$

Since we have the identity matrix on the left hand side, the resulting matrix on the right hand side is the inverse of A. So $A^{-1} = \begin{bmatrix} 1 & 0 & 1/3 \\ -1 & 1/2 & -1/2 \\ 0 & 0 & 1/3 \end{bmatrix}$.

GAUSS-JORDAN ELIMINATION: ADDITIONAL EXAMPLE

If we can't transform the matrix on the left into I, then A is not invertible. Let's see an example of this.

Suppose $A = \begin{bmatrix} 2 & 1 & 4 \\ -1 & 1 & 1 \\ 0 & 4 & 8 \end{bmatrix}$.

Let's form the augmented matrix with A on the left and the identity matrix on the right.

$$\begin{bmatrix} 2 & 1 & 4 & : & 1 & 0 & 0 \\ -1 & 1 & 1 & : & 0 & 1 & 0 \\ 0 & 4 & 8 & : & 0 & 0 & 1 \end{bmatrix}$$

Let's start doing row operations. Let's do $2R2 + R1 \rightarrow R2$:

$$\begin{bmatrix} 2 & 1 & 4 & : & 1 & 0 & 0 \\ 0 & 3 & 6 & : & 1 & 2 & 0 \\ 0 & 4 & 8 & : & 0 & 0 & 1 \end{bmatrix}$$

Let's do $\frac{1}{2} R1 \rightarrow R1$:

$$\begin{bmatrix} 1 & 1/2 & 2 & : & 1/2 & 0 & 0 \\ 0 & 3 & 6 & : & 1 & 2 & 0 \\ 0 & 4 & 8 & : & 0 & 0 & 1 \end{bmatrix}$$

Let's do $\frac{1}{3} R2 \rightarrow R2$:

$$\begin{bmatrix} 1 & 1/2 & 2 & : & 1/2 & 0 & 0 \\ 0 & 1 & 2 & : & 1/3 & 2/3 & 0 \\ 0 & 4 & 8 & : & 0 & 0 & 1 \end{bmatrix}$$

Let's do $\frac{1}{4} R3 \rightarrow R3$:

$$\begin{bmatrix} 1 & 1/2 & 2 & : & 1/2 & 0 & 0 \\ 0 & 1 & 2 & : & 1/3 & 2/3 & 0 \\ 0 & 1 & 2 & : & 0 & 0 & 1/4 \end{bmatrix}$$

Let's do $R2 - R3 \rightarrow R3$:

$$\begin{bmatrix} 1 & 1/2 & 2 & : & 1/2 & 0 & 0 \\ 0 & 1 & 2 & : & 1/3 & 2/3 & 0 \\ 0 & 0 & 0 & : & 1/3 & 2/3 & -1/4 \end{bmatrix}$$

Since we get a row of zeroes on the left hand side, we can't transform the left hand side to the identity matrix. Therefore, A is not invertible. That is, A does not have an inverse.

PROBLEM SET: INVERSE OF A MATRIX

For problems 1-5, use Gauss-Jordan elimination to find the inverse of A.

1. $A = \begin{bmatrix} 8 & -1 \\ 0 & 3 \end{bmatrix}$

2. $A = \begin{bmatrix} 1 & -1 \\ -1 & 0 \end{bmatrix}$

3. $A = \begin{bmatrix} 2 & 4 & 1 \\ 0 & 0 & 5 \\ -6 & 6 & 2 \end{bmatrix}$

4. $A = \begin{bmatrix} 1 & -1 & 0 \\ 2 & 2 & 3 \\ 5 & 2 & 1 \end{bmatrix}$

5. $A = \begin{bmatrix} 1 & -1 & 0 \\ 2 & 2 & 3 \\ 4 & 0 & 3 \end{bmatrix}$

6. Use the formula $\frac{1}{ad-bc}\begin{bmatrix} d & -b \\ -c & a \end{bmatrix}$ to find the inverse of A.

   $A = \begin{bmatrix} 4 & -1 \\ 8 & 3 \end{bmatrix}$

   Show that the resulting matrix B satisfies AB=BA=I.

SOLUTION SET: INVERSE OF A MATRIX

For problems 1-5, use Gauss-Jordan elimination to find the inverse of A.

1. $A = \begin{bmatrix} 8 & -1 \\ 0 & 3 \end{bmatrix}$

    Form the augmented matrix by adjoining the identity matrix:

    $\begin{bmatrix} 8 & -1 & 1 & 0 \\ 0 & 3 & 0 & 1 \end{bmatrix}$ (1/8)$R_1 \to R_1$

    $\begin{bmatrix} 1 & -\frac{1}{8} & \frac{1}{8} & 0 \\ 0 & 3 & 0 & 1 \end{bmatrix}$ (1/3)$R_2 \to R_2$

    $\begin{bmatrix} 1 & -\frac{1}{8} & \frac{1}{8} & 0 \\ 0 & 1 & 0 & \frac{1}{3} \end{bmatrix}$ (1/8)$R_2 + R_1 \to R_1$

    $\begin{bmatrix} 1 & 0 & \frac{1}{8} & \frac{1}{24} \\ 0 & 1 & 0 & \frac{1}{3} \end{bmatrix}$

    The left hand side is the identity matrix. The right hand side is the inverse of A. $A^{-1} = \begin{bmatrix} \frac{1}{8} & \frac{1}{24} \\ 0 & \frac{1}{3} \end{bmatrix}$.

2. $A = \begin{bmatrix} 1 & -1 \\ -1 & 0 \end{bmatrix}$

    Form the augmented matrix by adjoining the identity matrix:

    $\begin{bmatrix} 1 & -1 & 1 & 0 \\ -1 & 0 & 0 & 1 \end{bmatrix}$ $R_1 \leftrightarrow R_2$

    $\begin{bmatrix} -1 & 0 & 0 & 1 \\ 1 & -1 & 1 & 0 \end{bmatrix}$ $R_1 + R_2 \to R_2$

    $\begin{bmatrix} -1 & 0 & 0 & 1 \\ 0 & -1 & 1 & 1 \end{bmatrix}$ (-1)$R_1 \to R_1$

    $\begin{bmatrix} 1 & 0 & 0 & -1 \\ 0 & -1 & 1 & 1 \end{bmatrix}$ (-1)$R_2 \to R_2$

    $\begin{bmatrix} 1 & 0 & 0 & -1 \\ 0 & 1 & -1 & -1 \end{bmatrix}$

    The left hand side is now the identity matrix, and the right hand side is the inverse of A.

    $A^{-1} = \begin{bmatrix} 0 & -1 \\ -1 & -1 \end{bmatrix}$

3. $A = \begin{bmatrix} 2 & 4 & 1 \\ 0 & 0 & 5 \\ -6 & 6 & 2 \end{bmatrix}$

    Form the augmented matrix by adjoining the identity matrix.

$$\begin{bmatrix} 2 & 4 & 1 & 1 & 0 & 0 \\ 0 & 0 & 5 & 0 & 1 & 0 \\ -6 & 6 & 2 & 0 & 0 & 1 \end{bmatrix} \quad R_2 \leftrightarrow R_3$$

$$\begin{bmatrix} 2 & 4 & 1 & 1 & 0 & 0 \\ -6 & 6 & 2 & 0 & 0 & 1 \\ 0 & 0 & 5 & 0 & 1 & 0 \end{bmatrix} \quad (1/5)R_3 \to R_3$$

$$\begin{bmatrix} 2 & 4 & 1 & 1 & 0 & 0 \\ -6 & 6 & 2 & 0 & 0 & 1 \\ 0 & 0 & 1 & 0 & \frac{1}{5} & 0 \end{bmatrix} \quad 3R_1+R_2 \to R_2$$

$$\begin{bmatrix} 2 & 4 & 1 & 1 & 0 & 0 \\ 0 & 18 & 5 & 3 & 0 & 1 \\ 0 & 0 & 1 & 0 & \frac{1}{5} & 0 \end{bmatrix} \quad (1/18)R_2 \to R_2$$

$$\begin{bmatrix} 2 & 4 & 1 & 1 & 0 & 0 \\ 0 & 1 & \frac{5}{18} & \frac{1}{6} & 0 & \frac{1}{18} \\ 0 & 0 & 1 & 0 & \frac{1}{5} & 0 \end{bmatrix} \quad (-4)R_2+R_1 \to R_1$$

$$\begin{bmatrix} 2 & 0 & -\frac{1}{9} & \frac{1}{3} & 0 & -\frac{2}{9} \\ 0 & 1 & \frac{5}{18} & \frac{1}{6} & 0 & \frac{1}{18} \\ 0 & 0 & 1 & 0 & \frac{1}{5} & 0 \end{bmatrix} \quad (1/2)R_1 \to R_1$$

$$\begin{bmatrix} 1 & 0 & -\frac{1}{18} & \frac{1}{6} & 0 & -\frac{1}{9} \\ 0 & 1 & \frac{5}{18} & \frac{1}{6} & 0 & \frac{1}{18} \\ 0 & 0 & 1 & 0 & \frac{1}{5} & 0 \end{bmatrix} \quad (-5/18)R_3+R_2 \to R_2$$

$$\begin{bmatrix} 1 & 0 & -\frac{1}{18} & \frac{1}{6} & 0 & -\frac{1}{9} \\ 0 & 1 & 0 & \frac{1}{6} & -\frac{1}{18} & \frac{1}{18} \\ 0 & 0 & 1 & 0 & \frac{1}{5} & 0 \end{bmatrix} \quad (1/18)R_3+R_1 \to R_1$$

$$\begin{bmatrix} 1 & 0 & 0 & \frac{1}{6} & \frac{1}{90} & -\frac{1}{9} \\ 0 & 1 & 0 & \frac{1}{6} & -\frac{1}{18} & \frac{1}{18} \\ 0 & 0 & 1 & 0 & \frac{1}{5} & 0 \end{bmatrix}$$

The left hand side is the identity matrix, and the right hand side is the inverse of A.

$$A^{-1} = \begin{bmatrix} \frac{1}{6} & \frac{1}{90} & -\frac{1}{9} \\ \frac{1}{6} & -\frac{1}{18} & \frac{1}{18} \\ 0 & \frac{1}{5} & 0 \end{bmatrix}$$

4. $A = \begin{bmatrix} 1 & -1 & 0 \\ 2 & 2 & 3 \\ 5 & 2 & 1 \end{bmatrix}$

Form the augmented matrix by adjoining the identity matrix.

$$\begin{bmatrix} 1 & -1 & 0 & 1 & 0 & 0 \\ 2 & 2 & 3 & 0 & 1 & 0 \\ 5 & 2 & 1 & 0 & 0 & 1 \end{bmatrix} \quad -2R_1+R_2 \to R_2$$

$$\begin{bmatrix} 1 & -1 & 0 & 1 & 0 & 0 \\ 0 & 4 & 3 & -2 & 1 & 0 \\ 5 & 2 & 1 & 0 & 0 & 1 \end{bmatrix} \quad -5R_1+R_3 \to R_3$$

$$\begin{bmatrix} 1 & -1 & 0 & 1 & 0 & 0 \\ 0 & 4 & 3 & -2 & 1 & 0 \\ 0 & 7 & 1 & -5 & 0 & 1 \end{bmatrix} \quad (1/4)R_2 \to R_2$$

$$\begin{bmatrix} 1 & -1 & 0 & 1 & 0 & 0 \\ 0 & 1 & \frac{3}{4} & -\frac{1}{2} & \frac{1}{4} & 0 \\ 0 & 7 & 1 & -5 & 0 & 1 \end{bmatrix} \quad -7R_2+R_3 \to R_3$$

$$\begin{bmatrix} 1 & -1 & 0 & 1 & 0 & 0 \\ 0 & 1 & \frac{3}{4} & -\frac{1}{2} & \frac{1}{4} & 0 \\ 0 & 0 & -\frac{17}{4} & -\frac{3}{2} & -\frac{7}{4} & 1 \end{bmatrix} \quad (-4/17)R_3 \to R_3$$

$$\begin{bmatrix} 1 & -1 & 0 & 1 & 0 & 0 \\ 0 & 1 & \frac{3}{4} & -\frac{1}{2} & \frac{1}{4} & 0 \\ 0 & 0 & 1 & \frac{6}{17} & \frac{7}{17} & -\frac{4}{17} \end{bmatrix} \quad (-3/4)R_3+R_2 \to R_2$$

$$\begin{bmatrix} 1 & -1 & 0 & 1 & 0 & 0 \\ 0 & 1 & 0 & -\frac{13}{17} & -\frac{1}{17} & \frac{3}{17} \\ 0 & 0 & 1 & \frac{6}{17} & \frac{7}{17} & -\frac{4}{17} \end{bmatrix} \quad R_2+R_1 \to R_1$$

$$\begin{bmatrix} 1 & 0 & 0 & \frac{4}{17} & -\frac{1}{17} & \frac{3}{17} \\ 0 & 1 & 0 & -\frac{13}{17} & -\frac{1}{17} & \frac{3}{17} \\ 0 & 0 & 1 & \frac{6}{17} & \frac{7}{17} & -\frac{4}{17} \end{bmatrix}$$

The left hand side is the identity matrix, and the right hand side is the inverse of A.

$$A^{-1} = \begin{bmatrix} \frac{4}{17} & -\frac{1}{17} & \frac{3}{17} \\ -\frac{13}{17} & -\frac{1}{17} & \frac{3}{17} \\ \frac{6}{17} & \frac{7}{17} & -\frac{4}{17} \end{bmatrix}$$

5. $A = \begin{bmatrix} 1 & -1 & 0 \\ 2 & 2 & 3 \\ 4 & 0 & 3 \end{bmatrix}$

Form the augmented matrix by adjoining the identity matrix.

$$\begin{bmatrix} 1 & -1 & 0 & 1 & 0 & 0 \\ 2 & 2 & 3 & 0 & 1 & 0 \\ 4 & 0 & 3 & 0 & 0 & 1 \end{bmatrix} \quad -2R_1+R_2 \to R_2$$

$$\begin{bmatrix} 1 & -1 & 0 & 1 & 0 & 0 \\ 0 & 4 & 3 & -2 & 1 & 0 \\ 4 & 0 & 3 & 0 & 0 & 1 \end{bmatrix} \quad -4R_1+R_3 \to R_3$$

$$\begin{bmatrix} 1 & -1 & 0 & 1 & 0 & 0 \\ 0 & 4 & 3 & -2 & 1 & 0 \\ 0 & 4 & 3 & -4 & 0 & 1 \end{bmatrix} \quad R_2-R_3 \to R_3$$

$$\begin{bmatrix} 1 & -1 & 0 & 1 & 0 & 0 \\ 0 & 4 & 3 & -2 & 1 & 0 \\ 0 & 0 & 0 & 2 & 0 & -1 \end{bmatrix}$$

Note that we end up with a row of zeroes. So, A is not invertible.

6. Use the formula $\frac{1}{ad-bc}\begin{bmatrix} d & -b \\ -c & a \end{bmatrix}$ to find the inverse of A.

$$A = \begin{bmatrix} 4 & -1 \\ 8 & 3 \end{bmatrix}$$

Show that the resulting matrix B satisfies AB=BA=I.

The determinant ad-bc=4*3-(-1)*8=12+8=20.

$$\frac{1}{ad-bc}\begin{bmatrix} d & -b \\ -c & a \end{bmatrix} = \frac{1}{20}\begin{bmatrix} 3 & 1 \\ -8 & 4 \end{bmatrix} = \begin{bmatrix} \frac{3}{20} & \frac{1}{20} \\ -\frac{2}{5} & \frac{1}{5} \end{bmatrix}$$

$$\begin{bmatrix} 4 & -1 \\ 8 & 3 \end{bmatrix} \begin{bmatrix} \frac{3}{20} & \frac{1}{20} \\ -\frac{2}{5} & \frac{1}{5} \end{bmatrix} = \begin{bmatrix} 4*\frac{3}{20}+\frac{2}{5} & 4*\frac{1}{20}-\frac{1}{5} \\ 8*\frac{3}{20}-3*\frac{2}{5} & 8*\frac{1}{20}+3*\frac{1}{5} \end{bmatrix} = \begin{bmatrix} 1 & 0 \\ 0 & 1 \end{bmatrix}$$

$$\begin{bmatrix} \frac{3}{20} & \frac{1}{20} \\ -\frac{2}{5} & \frac{1}{5} \end{bmatrix} \begin{bmatrix} 4 & -1 \\ 8 & 3 \end{bmatrix} = \begin{bmatrix} \frac{3}{20}*4+\frac{1}{20}*8 & -\frac{3}{20}+\frac{3}{20} \\ -\frac{2}{5}*4+\frac{1}{5}*8 & \frac{2}{5}+\frac{3}{5} \end{bmatrix} = \begin{bmatrix} 1 & 0 \\ 0 & 1 \end{bmatrix}$$

SUMMARY: INVERSE OF A MATRIX

- The *inverse* of a matrix A is any matrix B such that when you multiply A on the left and right by the matrix B, you get the identity matrix.

- To find the inverse of a matrix, we can use a process called **Gauss-Jordan elimination**. We take the matrix A and adjoin the identity matrix. Then perform row operations on the resulting matrix until we transform A into I.

- If we can't transform the matrix on the left into I, then A is not invertible.

7 – DETERMINANTS

DETERMINANT OF A 2 BY 2 MATRIX

In this section, we're going to learn about determinants. The determinant of a 2×2 matrix $A = \begin{bmatrix} a & b \\ c & d \end{bmatrix}$ is defined as $ad - bc$ and is denoted $\det(A)$ and $|A|$. Let's do some examples.

Suppose $A = \begin{bmatrix} 2 & -1 \\ 3 & 4 \end{bmatrix}$. Let's find the determinant of A.

$\det(A) = ad - bc = 8 - (-3) = 11$.

Let's do another example.

Suppose $B = \begin{bmatrix} -1 & 2 \\ 3 & -2 \end{bmatrix}$.

Then, $\det(B) = ad - bc = (-1)(-2) - 3 \cdot 2 = 2 - 6 = -4$.

COFACTOR EXPANSION

To find the determinant of a 3×3 matrix or of a larger matrix, we have to use what's called ***cofactor expansion***. Let's do an example.

Suppose $A = \begin{bmatrix} 1 & 0 & -1 \\ 2 & 1 & 3 \\ -1 & 0 & 4 \end{bmatrix}$. First, we assign plus and minus signs to each position in the matrix. Start with the first position and assign a $+$ sign, and alternate the signs from left to right and from top to bottom like this:

$$\begin{bmatrix} + & - & + \\ - & + & - \\ + & - & + \end{bmatrix}$$

We want to expand along the first row. Take the first entry in the first column, 1. Multiply that by the sign in its position, $+$. Then, multiply by the determinant of the matrix that you get when you delete the first column and first row.

$$1 \cdot (+1) \cdot \begin{vmatrix} 1 & 3 \\ 0 & 4 \end{vmatrix}$$

Now, we move on to the second element in the first row, 0. Multiply that by the sign in its position, $-$. Then, multiply by the determinant of the matrix that you get when you delete the second column and first row.

$$0 \cdot (-1) \cdot \begin{vmatrix} 2 & 3 \\ -1 & 4 \end{vmatrix}$$

Now, move on to the third element in the first row, -1. Multiply that by the sign in its position, $+$. Then, multiply by the determinant of the matrix that you get when you delete the third column and first row.

$$(-1) \cdot (+1) \cdot \begin{vmatrix} 2 & 1 \\ -1 & 0 \end{vmatrix}$$

Finally, we add up the three expressions we got:

$$1 \cdot (+1) \cdot \begin{vmatrix} 1 & 3 \\ 0 & 4 \end{vmatrix} + 0 \cdot (-1) \cdot \begin{vmatrix} 2 & 3 \\ -1 & 4 \end{vmatrix} + (-1) \cdot (+1) \cdot \begin{vmatrix} 2 & 1 \\ -1 & 0 \end{vmatrix}$$

The expression $\begin{vmatrix} 1 & 3 \\ 0 & 4 \end{vmatrix}$ for the first term a_{11} is labeled M_{11} and is called the minor of a_{11}. The expression $\begin{vmatrix} 2 & 3 \\ -1 & 4 \end{vmatrix}$ for the second term a_{12} is labeled M_{12} and is the minor of a_{12}. The expression $\begin{vmatrix} 2 & 1 \\ -1 & 0 \end{vmatrix}$ is the minor of a_{13} and is labeled M_{13}. In general, M_{ij} is the **minor** of a_{ij} and is defined as the determinant of the matrix resulting from deleting the ith row and jth column.

The expression $(+1) \cdot \begin{vmatrix} 1 & 3 \\ 0 & 4 \end{vmatrix}$ for the first term a_{11} is called a cofactor and is labeled C_{11}. The expression $(-1) \cdot \begin{vmatrix} 2 & 3 \\ -1 & 4 \end{vmatrix}$ for the second term a_{12} is the cofactor for a_{12} and is labeled C_{12}. The expression $(+1) \cdot \begin{vmatrix} 2 & 1 \\ -1 & 0 \end{vmatrix}$ for the third term a_{13} is the cofactor for a_{13} and is labeled C_{13}. In general, C_{ij} is called a **cofactor** and is defined as $(-1)^{i+j} M_{ij}$.

So the det(A) is the sum of the terms a_{ij} multiplied by their corresponding cofactors C_{ij}, as we expand along the first row. You can expand along any row and the result is the same.

Ok, let's simplify what we had:

$$1 \cdot (+1) \cdot \begin{vmatrix} 1 & 3 \\ 0 & 4 \end{vmatrix} + 0 \cdot (-1) \cdot \begin{vmatrix} 2 & 3 \\ -1 & 4 \end{vmatrix} + (-1) \cdot (+1) \cdot \begin{vmatrix} 2 & 1 \\ -1 & 0 \end{vmatrix}$$

$$= 4 + 0 - 1$$

$$= 3$$

So det(A) = 3.

COFACTOR EXPANSION: ADDITIONAL EXAMPLES

Let's expand along the second row and see if we get the same result for the determinant of A.

Recall $A = \begin{bmatrix} 1 & 0 & -1 \\ 2 & 1 & 3 \\ -1 & 0 & 4 \end{bmatrix}$ and the signs are given by $\begin{bmatrix} + & - & + \\ - & + & - \\ + & - & + \end{bmatrix}$.

The first term in the second row is 2, and its sign is $-$. If we delete the second row and first column and take the determinant, the minor of a_{21} is given by $\begin{vmatrix} 0 & -1 \\ 0 & 4 \end{vmatrix}$.

So, we get $2(-1)\begin{vmatrix} 0 & -1 \\ 0 & 4 \end{vmatrix}$ for the first term.

The second term in the second row is 1, and its sign is $+$. If we delete the second row and second column and take the determinant, the minor of a_{22} is given by $\begin{vmatrix} 1 & -1 \\ -1 & 4 \end{vmatrix}$.

So, for the second term, 1, we get $1 \cdot (+1)\begin{vmatrix} 1 & -1 \\ -1 & 4 \end{vmatrix}$.

The third term in the second row is 3, and its sign is $-$. If we delete the second row and third column and take the determinant, the minor of a_{23} is given by $\begin{vmatrix} 1 & 0 \\ -1 & 0 \end{vmatrix}$.

So, for the third term, 3, we get $3 \cdot (-1)\begin{vmatrix} 1 & 0 \\ -1 & 0 \end{vmatrix}$.

Adding the three results, we get

$2(-1)\begin{vmatrix} 0 & -1 \\ 0 & 4 \end{vmatrix} + 1 \cdot (+1)\begin{vmatrix} 1 & -1 \\ -1 & 4 \end{vmatrix} + 3 \cdot (-1)\begin{vmatrix} 1 & 0 \\ -1 & 0 \end{vmatrix}$

$= -2 \cdot 0 + 1(4-1) - 3(0)$

$= 0 + 3 + 0$

$= 3$

This is the same result we got earlier by expanding along the first row. We can also expand along any column and the result is the same. Let's do an example.

Let's expand along the third column.

Recall $A = \begin{bmatrix} 1 & 0 & -1 \\ 2 & 1 & 3 \\ -1 & 0 & 4 \end{bmatrix}$ and the signs are given by $\begin{bmatrix} + & - & + \\ - & + & - \\ + & - & + \end{bmatrix}$.

Expanding along the third column, we get

$(-1) \cdot 1 \begin{vmatrix} 2 & 1 \\ -1 & 0 \end{vmatrix} + 3 \cdot (-1) \begin{vmatrix} 1 & 0 \\ -1 & 0 \end{vmatrix} + 4 \cdot (+1) \begin{vmatrix} 1 & 0 \\ 2 & 1 \end{vmatrix}$

$= (-1)(1) + 0 + 4(1)$

$= -1 + 4$

$= 3$

We get the same result as the earlier cofactor expansions.

PROBLEM SET: DETERMINANTS

Find the determinant of the 2 by 2 matrix.

1. $A = \begin{bmatrix} -1 & 1 \\ 0 & 1 \end{bmatrix}$

2. $A = \begin{bmatrix} 8 & 6 \\ 7 & -10 \end{bmatrix}$

Find the determinant using cofactor expansion.

1. $A = \begin{bmatrix} -1 & 0 & 1 \\ 2 & -1 & 0 \\ 3 & 4 & 6 \end{bmatrix}$

2. $A = \begin{bmatrix} -1 & 0 & 1 \\ 2 & -1 & 1 \\ 3 & 4 & 7 \end{bmatrix}$

3. $A = \begin{bmatrix} 1 & 0 & 0 & 4 \\ -1 & 2 & 3 & 6 \\ 1 & 0 & 1 & -1 \\ 4 & 4 & 1 & 0 \end{bmatrix}$

SOLUTION SET: DETERMINANTS

Find the determinant of the 2 by 2 matrix.

1. $A = \begin{bmatrix} -1 & 1 \\ 0 & 1 \end{bmatrix}$

   Use ad-bc. In this case, a=-1, b=1, c=0, d=1. So ad-bc=(-1)*1-0*1= -1.

2. $A = \begin{bmatrix} 8 & 6 \\ 7 & -10 \end{bmatrix}$

   Here, a=8, b=6, c=7, d= -10. So ad-bc=8*(-10)-6*7=-80-42= -122.

Find the determinant using cofactor expansion.

1. $A = \begin{bmatrix} -1 & 0 & 1 \\ 2 & -1 & 0 \\ 3 & 4 & 6 \end{bmatrix}$

   Expand along the first row.

   $(-1)\begin{vmatrix} -1 & 0 \\ 4 & 6 \end{vmatrix} + 1*\begin{vmatrix} 2 & -1 \\ 3 & 4 \end{vmatrix}$

   $=(-1)*(-6)+(8+3)$

   $=6+11=17$

2. $A = \begin{bmatrix} -1 & 0 & 1 \\ 2 & -1 & 1 \\ 3 & 4 & 7 \end{bmatrix}$

   Expand along the first row.

   $(-1)*\begin{vmatrix} -1 & 1 \\ 4 & 7 \end{vmatrix} + 1*\begin{vmatrix} 2 & -1 \\ 3 & 4 \end{vmatrix}$

   $=(-1)*(-11)+11$

   $=22$

3. $A = \begin{bmatrix} 1 & 0 & 0 & 4 \\ -1 & 2 & 3 & 6 \\ 1 & 0 & 1 & -1 \\ 4 & 4 & 1 & 0 \end{bmatrix}$

   Expand along the first row.

   $1*\begin{vmatrix} 2 & 3 & 6 \\ 0 & 1 & -1 \\ 4 & 1 & 0 \end{vmatrix} + (-1)*4*\begin{vmatrix} -1 & 2 & 3 \\ 1 & 0 & 1 \\ 4 & 4 & 1 \end{vmatrix}$

From here, we want to find $\begin{vmatrix} 2 & 3 & 6 \\ 0 & 1 & -1 \\ 4 & 1 & 0 \end{vmatrix}$ and $\begin{vmatrix} -1 & 2 & 3 \\ 1 & 0 & 1 \\ 4 & 4 & 1 \end{vmatrix}$ separately.

Starting with $\begin{vmatrix} 2 & 3 & 6 \\ 0 & 1 & -1 \\ 4 & 1 & 0 \end{vmatrix}$, expand along the first column.

$2*\begin{vmatrix} 1 & -1 \\ 1 & 0 \end{vmatrix}+4*\begin{vmatrix} 3 & 6 \\ 1 & -1 \end{vmatrix}=2*1+4*(-9)=2-36=-34$.

Next, find $\begin{vmatrix} -1 & 2 & 3 \\ 1 & 0 & 1 \\ 4 & 4 & 1 \end{vmatrix}$. Expand along second column.

$(-1)*2*\begin{vmatrix} 1 & 1 \\ 4 & 1 \end{vmatrix}+(-1)*4*\begin{vmatrix} -1 & 3 \\ 1 & 1 \end{vmatrix}=-2*(-3)-4*(-4)=6+16=22$.

Recall, we wanted to find $1*\begin{vmatrix} 2 & 3 & 6 \\ 0 & 1 & -1 \\ 4 & 1 & 0 \end{vmatrix}+(-1)*4*\begin{vmatrix} -1 & 2 & 3 \\ 1 & 0 & 1 \\ 4 & 4 & 1 \end{vmatrix}$.

This is equal to -34-4*22=-34-88=-122.

SUMMARY: DETERMINANTS

- The determinant of a 2 × 2 matrix $A = \begin{bmatrix} a & b \\ c & d \end{bmatrix}$ is defined as $ad - bc$ and is denoted $\det(A)$ and $|A|$.

- In general, M_{ij} is the **minor** of a_{ij} and is defined as the determinant of the matrix resulting from deleting the ith row and jth column.

- In general, C_{ij} is called a **cofactor** and is defined as $(-1)^{i+j} M_{ij}$.

- In **cofactor expansion**, the determinant is found by expanding along a column or row. It is given by $\sum_i a_{ij} (-1)^{i+j} M_{ij}$ if expanding along column j. It is given by $\sum_j a_{ij} (-1)^{i+j} M_{ij}$ if expanding along row i.

8 – PROPERTIES OF DETERMINANTS

DETERMINANT OF A PRODUCT OF MATRICES AND OF A SCALAR MULTIPLE OF A MATRIX

Let's look at properties of determinants. The first property states:

If A and B are $n \times n$ matrices, then $\det(AB) = \det(A)\det(B)$.

For example, suppose $A = \begin{bmatrix} 1 & 0 \\ -1 & 2 \end{bmatrix}$ and $B = \begin{bmatrix} 3 & 4 \\ 5 & 2 \end{bmatrix}$. Then $\det(A) = 2$ and $\det(B) = -14$. Taking the product, we get $\det(A)\det(B) = 2(-14) = -28$. Now, let's take the product AB and find its determinant to see if we get the same result.

$$AB = \begin{bmatrix} 1 & 0 \\ -1 & 2 \end{bmatrix}\begin{bmatrix} 3 & 4 \\ 5 & 2 \end{bmatrix} = \begin{bmatrix} 3 & 4 \\ 7 & 0 \end{bmatrix}$$

$$\det(AB) = -28$$

So we get the same result.

The next property of determinants states:

If c is a scalar and A is an $n \times n$ matrix, then $\det(cA) = c^n \det(A)$. Let's do an example.

Suppose $c = 8$ and $A = \begin{bmatrix} 1 & -1 \\ 3 & -4 \end{bmatrix}$.

$$cA = \begin{bmatrix} 8 & -8 \\ 24 & -32 \end{bmatrix}$$

So $\det(cA) = 8(-32) - (24)(-8) = -256 + 192 = -64$.

$n = 2$ since A is a 2×2 matrix. So $c^n = 8^2 = 64$.

$$\det(A) = -4 + 3 = -1$$

So $c^n \det(A) = -64$, the same result we got for $\det(cA)$.

Let's do another example.

Suppose $c = 3$ and $A = \begin{bmatrix} 1 & 0 & -1 \\ 2 & 0 & 3 \\ 1 & 4 & 2 \end{bmatrix}$.

$$cA = \begin{bmatrix} 3 & 0 & -3 \\ 6 & 0 & 9 \\ 3 & 12 & 6 \end{bmatrix}$$

Expanding along the second column, use cofactor expansion to find det(cA):

$$\det(cA) = 12(-1)\begin{vmatrix} 3 & -3 \\ 6 & 9 \end{vmatrix} = -12(27+18) = -12(45) = -540$$

Now, $n = 3$ since A is a 3×3 matrix. So $c^n = 3^3 = 27$. Expanding along the second column, use cofactor expansion to find det(A).

$$\det(A) = -4\begin{vmatrix} 1 & -1 \\ 2 & 3 \end{vmatrix} = -4(3+2) = -4(5) = -20$$

So $c^n \det(A) = 27(-20) = -540$, the same result we got for det(cA).

DETERMINANTS AND INVERTIBILITY

There is a nice connection between determinants and invertibility. A square matrix which is invertible has a nonzero determinant. Furthermore, if a square matrix has a nonzero determinant, then it's invertible. In other words, A is invertible iff $\det(A) \neq 0$. Let's do an example.

Determine if the matrix is invertible.

$$A = \begin{bmatrix} -2 & 3 & 1 \\ 4 & 2 & 1 \\ 2 & 5 & 2 \end{bmatrix}$$

Expand along the first row to find det(A).

$$\det(A) = (-2)\begin{vmatrix} 2 & 1 \\ 5 & 2 \end{vmatrix} + (3)(-1)\begin{vmatrix} 4 & 1 \\ 2 & 2 \end{vmatrix} + 1\begin{vmatrix} 4 & 2 \\ 2 & 5 \end{vmatrix}$$

$$= (-2)(-1) - 3(6) + 16$$

$$= 2 - 18 + 16$$

$$= 0$$

Since $\det(A) = 0$, A is not invertible.

Let's do another example.

Suppose $A = \begin{bmatrix} 1 & -1 & 9 \\ 0 & 2 & 2 \\ 1 & 3 & 2 \end{bmatrix}$.

Expanding along the first column to find det(A).

$$\det(A) = 1\begin{vmatrix} 2 & 2 \\ 3 & 2 \end{vmatrix} + 1\begin{vmatrix} -1 & 9 \\ 2 & 2 \end{vmatrix}$$

$$= -2 + (-20)$$

$$= -22$$

$\neq 0$

So A is invertible.

If we have an invertible matrix A, we can find the determinant of A^{-1} by using the formula $\det(A^{-1}) = \frac{1}{\det(A)}$. Let's do an example.

Suppose $A = \begin{bmatrix} 1 & -1 & 9 \\ 0 & 2 & 2 \\ 1 & 3 & 2 \end{bmatrix}$. We saw that the determinant of A is -22. So $\det(A^{-1}) = -\frac{1}{22}$.

DETERMINANT OF THE TRANSPOSE OF A MATRIX

The determinant of the transpose of a square matrix is the same as the determinant of the matrix.

$$\det(A^T) = \det(A)$$

For example, suppose $A = \begin{bmatrix} 1 & -1 & 9 \\ 0 & 2 & 2 \\ 1 & 3 & 2 \end{bmatrix}$. Then $A^T = \begin{bmatrix} 1 & 0 & 1 \\ -1 & 2 & 3 \\ 9 & 2 & 2 \end{bmatrix}$. To find the determinant of A^T, expand along the second column.

$$\det(A^T) = 2 \begin{vmatrix} 1 & 1 \\ 9 & 2 \end{vmatrix} - 2 \begin{vmatrix} 1 & 1 \\ -1 & 3 \end{vmatrix}$$

$$= 2(-7) - 2(4)$$

$$= -22$$

Recall that $\det(A) = -22$. So, $\det(A^T) = \det(A)$.

PROBLEM SET: PROPERTIES OF DETERMINANTS

1. Find det(AB) using the formula det(AB)=det(A)det(B).

   a. $A = \begin{bmatrix} 6 & -6 \\ 7 & 8 \end{bmatrix}$ $\qquad B = \begin{bmatrix} 100 & -1 \\ 0 & 10 \end{bmatrix}$

   b. $A = \begin{bmatrix} 1 & 6 & -7 \\ 0 & 1 & 2 \\ 0 & 3 & 4 \end{bmatrix}$ $\qquad B = \begin{bmatrix} -2 & -2 & 2 \\ 0 & 1 & 2 \\ 1 & -1 & 3 \end{bmatrix}$

2. Find det(cA) using the formula det(cA)=c^ndet(A).

   a. c=10 $\qquad A = \begin{bmatrix} 1 & -1 \\ 8 & 6 \end{bmatrix}$

   b. c=6 $\qquad A = \begin{bmatrix} 1 & 1 & 1 \\ 2 & 2 & 2 \\ 3 & 3 & 3 \end{bmatrix}$

3. Determine if the matrix is invertible using determinants.

   a. $A = \begin{bmatrix} 0 & 100 \\ 10 & 0 \end{bmatrix}$

   b. $A = \begin{bmatrix} 6 & 1 & 8 \\ -6 & 0 & -1 \\ 2 & 0 & 7 \end{bmatrix}$

   c. $A = \begin{bmatrix} 6 & 1 & 8 \\ -6 & 0 & -6 \\ 2 & 0 & 2 \end{bmatrix}$

SOLUTION SET: PROPERTIES OF DETERMINANTS

1. Find det(AB) using the formula det(AB)=det(A)det(B).

   a. $A = \begin{bmatrix} 6 & -6 \\ 7 & 8 \end{bmatrix}$ $B = \begin{bmatrix} 100 & -1 \\ 0 & 10 \end{bmatrix}$

   The det(A)=6*8-(-6)*7=90 and det(B)=1000. So, det(AB)=90*1000=90,000.

   b. $A = \begin{bmatrix} 1 & 6 & -7 \\ 0 & 1 & 2 \\ 0 & 3 & 4 \end{bmatrix}$ $B = \begin{bmatrix} -2 & -2 & 2 \\ 0 & 1 & 2 \\ 1 & -1 & 3 \end{bmatrix}$

   The det(A)=1*$\begin{vmatrix} 1 & 2 \\ 3 & 4 \end{vmatrix}$, by expanding along the first row. Then, we have det(A)=4-6=-2.

   The det(B)=(-2)*$\begin{vmatrix} 1 & 2 \\ -1 & 3 \end{vmatrix}$+1*$\begin{vmatrix} -2 & 2 \\ 1 & 2 \end{vmatrix}$, by expanding along the first column. Then, we have det(B)=(-2)*5+1*(-6)=-10-6=-16. So, det(AB)=det(A)*det(B)=(-2)*(-16)=32.

2. Find det(cA) using the formula det(cA)=c^ndet(A).

   a.   c=10 $A = \begin{bmatrix} 1 & -1 \\ 8 & 6 \end{bmatrix}$

   In this case, n=2. So $c^n=c^2=10^2=100$. Det(A)=14. We then have det(cA)=100*14=1400.

   b.   c=6 $A = \begin{bmatrix} 1 & 1 & 1 \\ 2 & 2 & 2 \\ 3 & 3 & 3 \end{bmatrix}$

   In this case, n=3. So $c^n=c^3=6^3=216$.

   Expanding along the first row, Det(A)=1*$\begin{vmatrix} 2 & 2 \\ 3 & 3 \end{vmatrix}$+(-1)*$\begin{vmatrix} 2 & 2 \\ 3 & 3 \end{vmatrix}$+1*$\begin{vmatrix} 2 & 2 \\ 3 & 3 \end{vmatrix}$=1*0-1*0+1*0=0. So det(cA)=216*0=0.

3. Determine if the matrix is invertible using determinants.

   a. $A = \begin{bmatrix} 0 & 100 \\ 10 & 0 \end{bmatrix}$

   Det(A)=-1000, which is not zero. Therefore, A is invertible.

   b. $A = \begin{bmatrix} 6 & 1 & 8 \\ -6 & 0 & -1 \\ 2 & 0 & 7 \end{bmatrix}$

   Expand along the second column. Det(A)=(-1)*$\begin{vmatrix} -6 & -1 \\ 2 & 7 \end{vmatrix}$. This simplifies to -1*(-40)=40, which is not zero. Therefore, A is invertible.

c. $A = \begin{bmatrix} 6 & 1 & 8 \\ -6 & 0 & -6 \\ 2 & 0 & 2 \end{bmatrix}$

Expand along the second column. Det(A)=1*(-1)$\begin{vmatrix} -6 & -6 \\ 2 & 2 \end{vmatrix}$. This simplifies to -1*0=0. Therefore, A is not invertible.

SUMMARY: PROPERTIES OF DETERMINANTS

- If A and B are $n \times n$ matrices, then $\det(AB) = \det(A)\det(B)$.

- If c is a scalar and A is an $n \times n$ matrix, then $\det(cA) = c^n \det(A)$.

- A is invertible iff $\det(A) \neq 0$.

- $\det(A^{-1}) = \dfrac{1}{\det(A)}$

- $\det(A^T) = \det(A)$

9 – VECTOR SPACES

VECTOR SPACE DEFINITION

In this section, we're going to look at vector spaces. A ***vector space*** is a set V, together with addition and scalar multiplication, such that the following ten properties hold:

Let $u, v, w \in V$ and $c, d \in \mathbb{R}$.

1. $u + v \in V$ closure under addition
2. $u + v = v + u$ commutativity under addition
3. $u + (v + w) = (u + v) + w$ associativity under addition
4. There is a zero vector $\mathbf{0}$ such that $u + \mathbf{0} = u$. Additive identity
5. For each $u \in V$ there is an additive inverse $-u$ such that $u + (-u) = \mathbf{0}$. Additive inverse
6. $cu \in V$ closure under scalar multiplication
7. $c(u + v) = cu + cv$ distributivity
8. $(c + d)u = cu + du$ distributivity
9. $c(du) = (cd)u$ associativity
10. $1u = u$ scalar identity

If any set V satisfies these ten properties, then it is a vector space.

VECTOR SPACE EXAMPLE

Let's look at an example of a vector space. $\mathbb{R}^2$ is a vector space. We're going to prove that $\mathbb{R}^2$ is a vector space.

Let $u = (u_1, u_2), v = (v_1, v_2), w = (w_1, w_2)$ lie in $\mathbb{R}^2$. Note that $u_1, u_2, v_1, v_2, w_1, w_2 \in \mathbb{R}$. Let's try to prove each of the vector space properties.

1. $u + v = (u_1, u_2) + (v_1, v_2)$

    $= (u_1 + v_1, u_2 + v_2) \in \mathbb{R}^2$ because $u_1 + v_1 \in \mathbb{R}$ and $u_2 + v_2 \in \mathbb{R}$ by closure of the real numbers.

2. $u + v = (u_1, u_2) + (v_1, v_2)$
    $= (u_1 + v_1, u_2 + v_2)$
    $= (v_1 + u_1, v_2 + u_2)$ by commutativity under addition of the real numbers
                     $(u_1 + v_1 = v_1 + u_1$ and $u_2 + v_2 = v_2 + u_2)$.
    $= (v_1, v_2) + (u_1, u_2)$
    $= v + u$

3. $u + (v + w) = (u_1, u_2) + ((v_1, v_2) + (w_1, w_2))$
    $= (u_1, u_2) + (v_1 + w_1, v_2 + w_2)$

    $= (u_1 + (v_1 + w_1), u_2 + (v_2 + w_2))$

$$= ((u_1 + v_1) + w_1, (u_2 + v_2) + w_2) \quad \text{by associativity under addition of the real numbers}$$

$$u_1 + (v_1 + w_1) = (u_1 + v_1) + w_1 \text{ and } u_2 + (v_2 + w_2) = (u_2 + v_2) + w_2$$

$$= (u_1 + v_1, u_2 + v_2) + (w_1, w_2)$$

$$= ((u_1, u_2) + (v_1, v_2)) + (w_1, w_2)$$

$$= (u + v) + w$$

4. $(0,0)$ is the additive identity.
$$u + (0,0) = (u_1 + 0, u_2 + 0)$$
$$= (u_1, u_2) \quad \text{because 0 is the additive identity for the real numbers.}$$
$$= u$$
So if we let $\mathbf{0} = (0,0)$, then $u + \mathbf{0} = u$. So there is an additive identity.

5. $-u$ is the additive inverse of u.
$$u + (-u) = (u_1, u_2) + (-u_1, -u_2)$$
$$= (u_1 - u_1, u_2 - u_2)$$
$$= (0,0) \quad \text{because the additive inverse of } u_1 \text{ is } -u_1 \text{ in } \mathbb{R}. \text{ Similarly, the additive inverse of } u_2 \text{ is } -u_2 \text{ in } \mathbb{R}.$$
$$= \mathbf{0}$$

6. $cu = c(u_1, u_2) = (cu_1, cu_2) \in \mathbb{R}^2 \quad$ since $cu_1, cu_2 \in \mathbb{R}$ by closure of multiplication in $\mathbb{R}$.

7. $c(u + v) = c((u_1, u_2) + (v_1, v_2))$
$$= c(u_1 + v_1, u_2 + v_2)$$
$$= (c(u_1 + v_1), c(u_2 + v_2))$$
$$= (cu_1 + cv_1, cu_2 + cv_2) \text{ by distributivity in } \mathbb{R}$$
$$= (cu_1, cu_2) + (cv_1, cv_2)$$
$$= c(u_1, u_2) + c(v_1, v_2)$$
$$= cu + cv$$

8. $(c + d)u = (c + d)(u_1, u_2)$
$$= ((c + d)u_1, (c + d)u_2)$$
$$= (cu_1 + du_1, cu_2 + du_2) \text{ by distributivity in } \mathbb{R}$$
$$= (cu_1, cu_2) + (du_1, du_2)$$
$$= c(u_1, u_2) + d(u_1, u_2)$$
$$= cu + du$$

9. $c(du) = c(d(u_1, u_2))$
$$= c(du_1, du_2)$$

$$= (c(du_1), c(du_2))$$

$$= ((cd)u_1, (cd)u_2) \quad \text{by associativity in } \mathbb{R}$$

$$= (cd)(u_1, u_2)$$

$$= (cd)u$$

10. $1u = 1(u_1, u_2)$
    $= (1u_1, 1u_2)$
    $= (u_1, u_2) \quad$ because 1 is the multiplicative identity in $\mathbb{R}$.
    $= u$

Since $\mathbb{R}^2$ satisfies the 10 properties of a vector space, $\mathbb{R}^2$ is a vector space. It's also true that $\mathbb{R}^n$ is a vector space for any $n > 2$. So $\mathbb{R}^3, \mathbb{R}^4$, etc. are all vector spaces.

VECTOR SPACE: ADDITIONAL EXAMPLE

The example of $\mathbb{R}^n$ is not the only example of vector space. Vector spaces can be very different in what they look like. For example, the set $M_{m,n}$ of all m by n matrices, with matrix addition and scalar multiplication, forms a vector space. The vectors in this case are matrices, which is a little strange; but it's fine, if it satisfies the properties of a vector space then it's a vector space. Another example is the set of all polynomials of degree less than or equal to n; this forms a vector space. In this case, the vectors are polynomials.

Let's do an example. Consider P_2, the set of all polynomials of degree less than or equal to 2. This is a vector space. We want to try to prove this. We would have to show all 10 properties of a vector space. Let's show the first 5 properties.

Let $f, g, h \in P_2$ and $c, d \in \mathbb{R}$. Then f, g, h will look like this:

$f(x) = a_2 x^2 + a_1 x + a_0 \quad$ where $a_0, a_1, a_2 \in \mathbb{R}$

$g(x) = b_2 x^2 + b_1 x + b_0 \quad$ where $b_0, b_1, b_2 \in \mathbb{R}$

$h(x) = c_2 x^2 + c_1 x + c_0 \quad$ where $c_0, c_1, c_2 \in \mathbb{R}$

1. $(f + g)(x) = f(x) + g(x)$ by definition
    $= (a_2 x^2 + a_1 x + a_0) + (b_2 x^2 + b_1 x + b_0)$

    $= (a_2 + b_2)x^2 + (a_1 + b_1)x + (a_0 + b_0)$, which is a polynomial of degree less than or equal to 2 and has real coefficients.

    $\Rightarrow \quad f + g \in P_2$

2. $(f + g)(x) = f(x) + g(x)$ by definition

$$= (a_2x^2 + a_1x + a_0) + (b_2x^2 + b_1x + b_0)$$
$$= (a_2 + b_2)x^2 + (a_1 + b_1)x + (a_0 + b_0)$$
$$= (b_2 + a_2)x^2 + (b_1 + a_1)x + (b_0 + a_0) \text{ by commutativity of } \mathbb{R}$$
$$= (b_2x^2 + b_1x + b_0) + (a_2x^2 + a_1x + a_0)$$
$$= g(x) + f(x)$$
$$= (g + f)(x)$$

$\Longrightarrow f + g = g + f.$

3. We want to show $f + (g + h) = (f + g) + h$.
$$(f + (g + h))(x) = f(x) + (g + h)(x)$$
$$= f(x) + ((g(x) + h(x))$$
$$= a_2x^2 + a_1x + a_0 + ((b_2x^2 + b_1x + b_0) + (c_2x^2 + c_1x + c_0))$$
$$= a_2x^2 + a_1x + a_0 + ((b_2 + c_2)x^2 + (b_1 + c_1)x + (b_0 + c_0))$$
$$= (a_2 + (b_2 + c_2))x^2 + (a_1 + (b_1 + c_1))x + (a_0 + (b_0 + c_0))$$
$$= ((a_2 + b_2) + c_2)x^2 + ((a_1 + b_1) + c_1)x + ((a_0 + b_0) + c_0) \quad \text{by associativity of real numbers}$$
$$= (a_2 + b_2)x^2 + (a_1 + b_1)x + (a_0 + b_0) + c_2x^2 + c_1x + c_0$$
$$= ((a_2x^2 + a_1x + a_0) + (b_2x^2 + b_1x + b_0)) + (c_2x^2 + c_1x + c_0)$$
$$= (f(x) + g(x)) + h(x)$$
$$= ((f + g)(x)) + h(x)$$
$$= ((f + g) + h)(x)$$

$\Longrightarrow (f + (g + h) = (f + g) + h$

4. Let $\mathbf{0}(x) = 0$.
Then $(\mathbf{0} + f)(x) = \mathbf{0}(x) + f(x)$
$$= 0 + (a_2x^2 + a_1x + a_0)$$
$$= a_2x^2 + a_1x + (a_0 + 0) \quad \text{since 0 is the additive identity in } \mathbb{R}$$
$$= a_2x^2 + a_1x + a_0$$
$$= f(x)$$

$\Longrightarrow \mathbf{0} + f = f$

5. The additive inverse of f is the polynomial $-f$ defined by $(-f)(x) = -f(x)$.

$$(f + (-f))(x) = f(x) + (-f)(x)$$
$$= f(x) - f(x)$$
$$= (a_2 x^2 + a_1 x + a_0) - (a_2 x^2 + a_1 x + a_0)$$
$$= (a_2 - a_2)x^2 + (a_1 - a_1)x + (a_0 - a_0)$$
$$= 0x^2 + 0x + 0$$
$$= 0$$
$$= \mathbf{0}(x)$$

$$\implies f + (-f) = \mathbf{0}$$

We've shown the first 5 properties of a vector space applied to P_2. Try proving the remaining 5 properties of a vector space for P_2 in the problem set.

PROBLEM SET: VECTOR SPACES

1. Prove the vector space properties 6-10 for P_2, the set of all polynomials of degree less than or equal to 2.

2. Prove that $M_{2,2}$, the set of all 2 by 2 matrices, is a vector space.

SOLUTION SET: VECTOR SPACES

1. Let $f, g, h \in P_2$ and $c, d \in \mathbb{R}$. Suppose
   $f(x) = a_2 x^2 + a_1 x + a_0$, where $a_0, a_1, a_2 \in \mathbb{R}$
   $g(x) = b_2 x^2 + b_1 x + b_0$, where $b_0, b_1, b_2 \in \mathbb{R}$
   $h(x) = c_2 x^2 + c_1 x + c_0$, where $c_0, c_1, c_2 \in \mathbb{R}$

   For condition 6 of the definition of vector space, we want to show $cf \in P_2$.
   $$(cf)(x) = cf(x) = c(a_2 x^2 + a_1 x + a_0) = (ca_2)x^2 + (ca_1)x + (ca_0)$$

   $$\Rightarrow cf \in P_2$$

   For condition 7, we need to show $c(f + g) = cf + cg$.
   $$\big(c(f+g)\big)(x) = c\big((f+g)(x)\big) = c\big(f(x) + g(x)\big) = cf(x) + cg(x) = (cf)(x) + (cg)(x)$$

   $$= (cf + cg)(x)$$

   $$\Rightarrow c(f + g) = cf + cg$$

   For condition 8, we need to show $(c + d)f = cf + df$.
   $$\big((c+d)f\big)(x) = (c+d)f(x)$$
   $$= cf(x) + df(x)$$
   $$= (cf)(x) + (df)(x)$$
   $$= (cf + df)(x)$$

   $$\Rightarrow (c + d)f = cf + df$$

   For condition 9, we need to show $c(df) = (cd)f$.
   $$\big(c(df)\big)(x) = c(df)(x) = c\big(df(x)\big)$$
   $$= (cd)f(x)$$
   $$= \big((cd)f\big)(x)$$

   $$\Rightarrow c(df) = (cd)f$$

   For condition 10, we need to show $1 \cdot f = f$.
   $$(1 \cdot f)(x) = 1 \cdot f(x) = f(x)$$

   $$\Rightarrow 1 \cdot f = f$$

2. $M_{2,2} = \left\{ \begin{bmatrix} a_{11} & a_{12} \\ a_{21} & a_{22} \end{bmatrix} \mid a_{11}, a_{12}, a_{21}, a_{22} \in \mathbb{R} \right\}$.

   Let $A = \begin{bmatrix} a_{11} & a_{12} \\ a_{21} & a_{22} \end{bmatrix}, B = \begin{bmatrix} b_{11} & b_{12} \\ b_{21} & b_{22} \end{bmatrix}, C = \begin{bmatrix} c_{11} & c_{12} \\ c_{21} & c_{22} \end{bmatrix}$ be in $M_{2,2}$ and $c, d \in \mathbb{R}$.

1. Show $A + B \in M_{2,2}$. $\quad A + B = \begin{bmatrix} a_{11} + b_{11} & a_{12} + b_{12} \\ a_{21} + b_{21} & a_{22} + b_{22} \end{bmatrix} \in M_{2,2}$.

2. Show $A + B = B + A$.

$$A + B = \begin{bmatrix} a_{11} + b_{11} & a_{12} + b_{12} \\ a_{21} + b_{21} & a_{22} + b_{22} \end{bmatrix}$$

$$= \begin{bmatrix} b_{11} + a_{11} & b_{12} + a_{12} \\ b_{21} + a_{21} & b_{22} + a_{22} \end{bmatrix} \quad \text{by commutativity of } \mathbb{R}$$

$$= B + A$$

3. Show $A + (B + C) = (A + B) + C$.

$$A + (B + C) = \begin{bmatrix} a_{11} & a_{12} \\ a_{21} & a_{22} \end{bmatrix} + \begin{bmatrix} b_{11} + c_{11} & b_{12} + c_{12} \\ b_{21} + c_{21} & b_{22} + c_{22} \end{bmatrix}$$

$$= \begin{bmatrix} a_{11} + (b_{11} + c_{11}) & a_{12} + (b_{12} + c_{12}) \\ a_{21} + (b_{21} + c_{21}) & a_{22} + (b_{22} + c_{22}) \end{bmatrix}$$

$$= \begin{bmatrix} (a_{11} + b_{11}) + c_{11} & (a_{12} + b_{12}) + c_{12} \\ (a_{21} + b_{21}) + c_{21} & (a_{22} + b_{22}) + c_{22} \end{bmatrix}$$

$$\text{by associativity of addition in } \mathbb{R}$$

$$= (A + B) + C$$

4. Let $\mathbf{0} = \begin{bmatrix} 0 & 0 \\ 0 & 0 \end{bmatrix}$. Then $A + \mathbf{0} = \begin{bmatrix} a_{11} & a_{12} \\ a_{21} & a_{22} \end{bmatrix} + \begin{bmatrix} 0 & 0 \\ 0 & 0 \end{bmatrix}$

$$= \begin{bmatrix} a_{11} + 0 & a_{12} + 0 \\ a_{21} + 0 & a_{22} + 0 \end{bmatrix}$$

$$= \begin{bmatrix} a_{11} & a_{12} \\ a_{21} & a_{22} \end{bmatrix}$$

$$= A$$

5. The inverse of A is $-A$.

$$A + (-A) = \begin{bmatrix} a_{11} & a_{12} \\ a_{21} & a_{22} \end{bmatrix} + \left(- \begin{bmatrix} a_{11} & a_{12} \\ a_{21} & a_{22} \end{bmatrix}\right)$$

$$= \begin{bmatrix} a_{11} & a_{12} \\ a_{21} & a_{22} \end{bmatrix} + \begin{bmatrix} -a_{11} & -a_{12} \\ -a_{21} & -a_{22} \end{bmatrix}$$

$$= \begin{bmatrix} a_{11} - a_{11} & a_{12} - a_{12} \\ a_{21} - a_{21} & a_{22} - a_{22} \end{bmatrix}$$

$$= \begin{bmatrix} 0 & 0 \\ 0 & 0 \end{bmatrix} = \mathbf{0}$$

6. Show $cA \in M_{2,2}$.

$$cA = c \begin{bmatrix} a_{11} & a_{12} \\ a_{21} & a_{22} \end{bmatrix} = \begin{bmatrix} ca_{11} & ca_{12} \\ ca_{21} & ca_{22} \end{bmatrix} \in M_{2,2}$$

7. Show $c(A+B) = cA + cB$.

$$c(A+B) = c\begin{bmatrix} a_{11}+b_{11} & a_{12}+b_{12} \\ a_{21}+b_{21} & a_{22}+b_{22} \end{bmatrix}$$

$$= \begin{bmatrix} c(a_{11}+b_{11}) & c(a_{12}+b_{12}) \\ c(a_{21}+b_{21}) & c(a_{22}+b_{22}) \end{bmatrix}$$

$$= \begin{bmatrix} ca_{11}+cb_{11} & ca_{12}+cb_{12} \\ ca_{21}+cb_{21} & ca_{22}+cb_{22} \end{bmatrix}$$

$$= \begin{bmatrix} ca_{11} & ca_{12} \\ ca_{21} & ca_{22} \end{bmatrix} + \begin{bmatrix} cb_{11} & cb_{12} \\ cb_{21} & cb_{22} \end{bmatrix}$$

$$= cA + cB$$

8. Show $(c+d)A = cA + dA$.

$$(c+d)A = (c+d)\begin{bmatrix} a_{11} & a_{12} \\ a_{21} & a_{22} \end{bmatrix}$$

$$= \begin{bmatrix} (c+d)a_{11} & (c+d)a_{12} \\ (c+d)a_{21} & (c+d)a_{22} \end{bmatrix}$$

$$= \begin{bmatrix} ca_{11}+da_{11} & ca_{12}+da_{12} \\ ca_{21}+da_{21} & ca_{22}+da_{22} \end{bmatrix}$$

$$= \begin{bmatrix} ca_{11} & ca_{12} \\ ca_{21} & ca_{22} \end{bmatrix} + \begin{bmatrix} da_{11} & da_{12} \\ da_{21} & da_{22} \end{bmatrix}$$

$$= cA + dA$$

9. Show $c(dA) = (cd)A$.

$$c(dA) = c\left(d\begin{bmatrix} a_{11} & a_{12} \\ a_{21} & a_{22} \end{bmatrix}\right) = c\begin{bmatrix} da_{11} & da_{12} \\ da_{21} & da_{22} \end{bmatrix}$$

$$= \begin{bmatrix} c(da_{11}) & c(da_{12}) \\ c(da_{21}) & c(da_{22}) \end{bmatrix}$$

$$= \begin{bmatrix} (cd)a_{11} & (cd)a_{12} \\ (cd)a_{21} & (cd)a_{22} \end{bmatrix} \text{ by associativity of multiplication in } \mathbb{R}$$

$$= (cd)\begin{bmatrix} a_{11} & a_{12} \\ a_{21} & a_{22} \end{bmatrix}$$

$$= (cd)A$$

10. Show $1 \cdot A = A$.

$$1 \cdot A = 1 \cdot \begin{bmatrix} a_{11} & a_{12} \\ a_{21} & a_{22} \end{bmatrix}$$

$$= \begin{bmatrix} 1 \cdot a_{11} & 1 \cdot a_{12} \\ 1 \cdot a_{21} & 1 \cdot a_{22} \end{bmatrix}$$

$$= \begin{bmatrix} a_{11} & a_{12} \\ a_{21} & a_{22} \end{bmatrix}$$

$$= A$$

Therefore, $M_{2,2}$ is a vector space.

EXAMPLES OF SETS THAT ARE NOT VECTOR SPACES

We're now going to look at examples of sets that are not vector spaces. Consider the set of polynomials that have degree exactly 2.

Let $f(x) = 3x^2 - x + 2$ and $g(x) = -3x^2 + 4x + 5$.

Then f, g are polynomials of degree exactly 2.

However, $(f + g)(x) = (3x^2 - x + 2) + (-3x^2 + 4x + 5)$

$$= 3x + 7$$

$\implies f + g$ is a polynomial of degree 1.

$\implies f + g$ is not in the set of polynomials of degree 2.

$\implies$ closure under addition does not hold.

Since at least one of the vector space properties fails, the set of polynomials that have degree exactly 2 is not a vector space.

Let's look at another example. Consider $\mathbb{Z}^2$, the set of all pairs of the form (m, n) where $m, n \in \mathbb{Z}$. $\mathbb{Z}^2$ is not a vector space.

Let $u = (2, 3)$ and $c = \frac{1}{3}$.

Then $cu = \frac{1}{3}(2, 3) = (\frac{2}{3}, 1)$, which does not lie in $\mathbb{Z}^2$ since $\frac{2}{3}$ does not lie in $\mathbb{Z}$.

$\implies cu$ does not lie in $\mathbb{Z}^2$

$\implies$ closure under scalar multiplication does not hold.

Since at least one of the vector space properties fails, the set $\mathbb{Z}^2$ is not a vector space.

PROBLEM SET: SETS THAT ARE NOT VECTOR SPACES

1. Prove that the set of all invertible 2 by 2 matrices, with matrix multiplication and scalar multiplication, is not a vector space.

2. Determine if the set $\{(x, -x) | x \in \mathbb{R}\}$, with the standard addition and scalar multiplication in $\mathbb{R}^2$, is a vector space. If so, prove all vector space properties. If not, identify which vector space properties fail.

3. Consider $\mathbb{R}^2$ with the following addition and scalar multiplication:
$$(x_1, y_1) + (x_2, y_2) = (x_1 y_1, x_2 y_2)$$
$$c(x_1, y_1) = (cx_1, cy_1).$$

Determine if $\mathbb{R}^2$, with the above addition and scalar multiplication, is a vector space. If so, prove all ten vector space properties. If not, identify which properties fail.

SOLUTION SET: SETS THAT ARE NOT VECTOR SPACES

1. Let $A = \begin{bmatrix} 1 & 8 \\ 7 & 2 \end{bmatrix}$ and $B = \begin{bmatrix} 1 & -8 \\ 7 & -2 \end{bmatrix}$.

   Then $\det(A) = -54 \neq 0$ and $\det(B) = 54 \neq 0$.

   $\Rightarrow A, B$ are invertible.

   However, $A + B = \begin{bmatrix} 2 & 0 \\ 14 & 0 \end{bmatrix}$ and $\det(A + B) = 0$.

   $\Rightarrow A + B$ is not invertible.

   $\Rightarrow A + B$ does not lie in the set of all invertible 2 by 2 matrices.

2. Yes, the set $V = \{(x, -x) | x \in \mathbb{R}\}$ is a vector space. Let's show all ten vector space properties. Let $(x, -x)$ and $(y, -y)$ lie in V, and let $c, d \in \mathbb{R}$.

   1. $(x, -x) + (y, -y) = (x + y, -x - y) = (x + y, -(x + y)) \in V$

   2. $(x, -x) + (y, -y) = (x + y, -x - y) = (y + x, -y - x) = (y, -y) + (x, -x)$.

   3. $(x, -x) + \big((y, -y) + (z, -z)\big) = (x, -x) + (y + z, -y - z)$
      $= \big(x + (y + z), -x + (-y - z)\big)$
      $= \big((x + y) + z, (-x - y) - z\big)$
      $= (x + y, -x - y) + (z, -z)$
      $= \big((x, -x) + (y, -y)\big) + (z, -z)$

   4. $\mathbf{0} = (0, 0) \in V$ and $(x, -x) + (0, 0) = (x + 0, -x + 0) = (x, -x)$

   5. The inverse of $(x, -x)$ is $(-x, x)$.

      $$(x, -x) + (-x, x) = (x - x, -x + x) = (0, 0)$$

   6. $c(x, -x) = \big(cx, c(-x)\big) = (cx, -cx) \in V$.

   7. $c\big((x, -x) + (y, -y)\big) = c(x + y, -(x + y))$

      $= \big(c(x + y), -c(x + y)\big)$
      $= (cx + cy, -cx - cy)$
      $= (cx, -cx) + (cy, -cy)$
      $= c(x, -x) + c(y, -y)$

8. $(c+d)(x,-x) = \bigl((c+d)x, -(c+d)x\bigr)$

$$= (cx+dx, -cx-dx)$$
$$= (cx,-cx) + (dx,-dx)$$
$$= c(x,-x) + d(x,-x)$$

9. $(cd)(x,-x) = \bigl((cd)x, -(cd)x\bigr) = \bigl(c(dx), -c(dx)\bigr)$

$$= \bigl(c(dx), c(-dx)\bigr)$$
$$= c(dx, -dx)$$
$$= c\bigl(dx, d(-x)\bigr)$$
$$= c(d(x,-x))$$

10. $1 \cdot (x,-x) = \bigl(1\cdot x, 1\cdot(-x)\bigr) = (x,-x)$

3. No, $\mathbb{R}^2$, with the given addition and scalar multiplication, is not a vector space. The commutativity property and both distributivity properties fail.

SUMMARY: VECTOR SPACES

- A ***vector space*** is a set V, together with addition and scalar multiplication, such that the following ten properties hold:

  Let $u, v, w \in V$ and $c, d \in \mathbb{R}$.

  1. $u + v \in V$ closure under addition
  2. $u + v = v + u$ commutativity under addition
  3. $u + (v + w) = (u + v) + w$ associativity under addition
  4. There is a zero vector $\mathbf{0}$ such that $u + \mathbf{0} = u$. Additive identity
  5. For each $u \in V$ there is an additive inverse $-u$ such that $u + (-u) = \mathbf{0}$. Additive inverse
  6. $cu \in V$ closure under scalar multiplication
  7. $c(u + v) = cu + cv$ distributivity
  8. $(c + d)u = cu + du$ distributivity
  9. $c(du) = (cd)u$ associativity
  10. $1u = u$ scalar identity

- $\mathbb{R}^n$ and P_n, the set of polynomials of degree less than or equal to n, are vector spaces.

- If a set fails to satisfy at least one vector space property, it does not form a vector space.

10 – SUBSPACES

SUBSPACE DEFINITION AND SUBSPACE PROPERTIES

In this section, we're going to look at subspaces.

A subset W of a vector space V is a **subspace** of V if W is nonempty and a vector space itself with the same operations as V. For example, the set $W = \{(x, 0) | x \in \mathbb{R}\}$ is a subspace of $V = \mathbb{R}^2$. $\mathbb{R}^2$ consists of all points in the x-y plane. W consists of all the points on the x-axis. The claim is that the x-axis is a subspace of the entire plane $\mathbb{R}^2$.

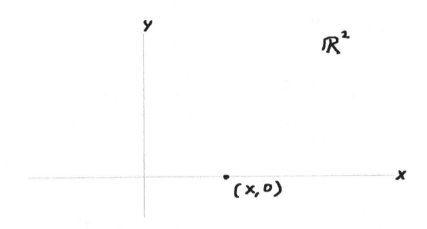

To show that W is a subspace of $\mathbb{R}^2$, we'd have to show that W has all ten vector space properties. Fortunately, it turns out that we only need to show two closure properties and that W is nonempty. To show that W is nonempty, we can show that it contains the zero vector. The subspace properties consist of:

1. $\mathbf{0} \in W$
2. $u + v \in W$ for all $u, v \in W$.   Closure under addition
3. $cu \in W$ for all $c \in \mathbb{R}, u \in W$.   Closure under scalar multiplication

Let's do an example.

Suppose $W = \{(x, 0) | x \in \mathbb{R}\}$ and $V = \mathbb{R}^2$. Let's show that W is a subspace of V.

1. Let $x = 0$. Then $(x, 0) = (0,0)$ and $0 \in \mathbb{R}$. So $\mathbf{0} \in W$.
2. Let $u = (x, 0)$ and $v = (y, 0)$ where $x, y \in \mathbb{R}$.
   Then $u + v = (x, 0) + (y, 0) = (x + y, 0) \in W$ because $x + y \in \mathbb{R}$ by closure of addition in $\mathbb{R}$.
3. Let $c \in \mathbb{R}$ and $u = (x, 0) \in W$.
   Then $cu = c(x, 0) = (cx, c \cdot 0) = (cx, 0) \in W$ because $cx \in \mathbb{R}$ by closure of multiplication in $\mathbb{R}$.

So W is a subspace of $\mathbb{R}^2$.

DEFINITION OF TRIVIAL AND NONTRIVIAL SUBSPACE

The subset consisting of only the zero vector $\{\mathbf{0}\}$ is a subspace of V. It's called the *zero subspace*. The set V itself is a subspace of V. These two subspaces are called *trivial subspaces* of V.

A *nontrivial subspace* is any subspace that is not the zero subspace or the whole vector space V. We've seen that $W = \{(x, 0) | x \in \mathbb{R}\}$ is a subspace that is not the zero subspace nor the entire vector space $\mathbb{R}^2$. W is not the zero subspace because it contains vectors other than the zero vector; for instance, it contains $(1,0)$. W is not the whole space $\mathbb{R}^2$ because it won't contain other vectors in $\mathbb{R}^2$; for instance, it doesn't contain $(2, 3)$. So W is a nontrivial subspace of $\mathbb{R}^2$.

ADDITIONAL EXAMPLE OF SUBSPACE

Let's look at another example of a subspace.

Consider the set W of all matrices of the form $\begin{bmatrix} a & 0 \\ b & c \end{bmatrix}$, where a, b, c are real numbers. This is a subset of $M_{2,2}$, the set of all 2×2 matrices. We want to show the three subspace properties for W.

1. Let $a = b = c = 0$. Then the zero matrix $\begin{bmatrix} 0 & 0 \\ 0 & 0 \end{bmatrix}$ lies in W. This is true because a, b, c are real numbers, and the entry in the first row and second column is 0.

2. Let $u = \begin{bmatrix} a & 0 \\ b & c \end{bmatrix}$ and $v = \begin{bmatrix} d & 0 \\ e & f \end{bmatrix}$ be in W. We want to show that $u + v \in W$.
$$u + v = \begin{bmatrix} a & 0 \\ b & c \end{bmatrix} + \begin{bmatrix} d & 0 \\ e & f \end{bmatrix} = \begin{bmatrix} a+d & 0 \\ b+e & c+f \end{bmatrix}$$
Since $a + d, b + e, c + f$ are real numbers and the entry in the first row and second column is 0, $\begin{bmatrix} a+d & 0 \\ b+e & c+f \end{bmatrix}$ lies in W. So $u + v \in W$.

3. Let $u = \begin{bmatrix} a & 0 \\ b & c \end{bmatrix}$ be in W and $k \in \mathbb{R}$.
Then $ku = k \begin{bmatrix} a & 0 \\ b & c \end{bmatrix} = \begin{bmatrix} ka & 0 \\ kb & kc \end{bmatrix}$.
Since ka, kb, kc are real numbers and the entry in the first row and second column is 0, $\begin{bmatrix} ka & 0 \\ kb & kc \end{bmatrix}$ lies in W. So $ku \in W$.

Since all three subspace properties hold for W, W is a subspace of $M_{2,2}$.

PROBLEM SET: SUBSPACES

1. Show that the set $W = \{(x, 2x) | x \in \mathbb{R}\}$ is a subspace of $\mathbb{R}^2$ using the subspace properties. Show that W is a nontrivial subspace.

2. Show that the set $W = \left\{ k \cdot \begin{bmatrix} 1 & -1 \\ 2 & 3 \end{bmatrix} \middle| k \in \mathbb{R} \right\}$ is a subspace of $M_{2,2}$. Show that W is a nontrivial subspace.

SOLUTION SET: SUBSPACES

1. $W = \{(x, 2x) | x \in \mathbb{R}\}$

   1. Let $x = 0$. Then $(x, 2x) = (0, 2(0)) = (0,0)$ and $(0,0) \in W$.

   2. Let $(x, 2x)$ and $(y, 2y)$ lie in W.
      Then $(x, 2x) + (y, 2y) = (x + y, 2x + 2y) = (x + y, 2(x + y)) \in W$

   3. Let $c \in \mathbb{R}$ and $(x, 2x) \in W$. Then $c(x, 2x) = (cx, c(2x)) = (cx, 2(cx)) \in W$

   To show that W is nontrivial, we need to show that W contains more than just the zero vector and also that there is some element in $\mathbb{R}^2$ that is not in W.

   $(1,2) \in W$

   $\Rightarrow W \neq \{(0,0)\}$

   $(2,7) \in \mathbb{R}^2$, but $(2,7)$ does not lie in W.

   $\Rightarrow W \neq \mathbb{R}^2$.

2. $W = \left\{ k \cdot \begin{bmatrix} 1 & -1 \\ 2 & 3 \end{bmatrix} \middle| k \in \mathbb{R} \right\}$

   1. Let $k = 0$. Then $0 \cdot \begin{bmatrix} 1 & -1 \\ 2 & 3 \end{bmatrix} = \begin{bmatrix} 0 & 0 \\ 0 & 0 \end{bmatrix}$. So $\begin{bmatrix} 0 & 0 \\ 0 & 0 \end{bmatrix} \in W$

   2. Let $k_1 \cdot \begin{bmatrix} 1 & -1 \\ 2 & 3 \end{bmatrix}$ and $k_2 \cdot \begin{bmatrix} 1 & -1 \\ 2 & 3 \end{bmatrix}$ lie in W.
      Then $k_1 \cdot \begin{bmatrix} 1 & -1 \\ 2 & 3 \end{bmatrix} + k_2 \cdot \begin{bmatrix} 1 & -1 \\ 2 & 3 \end{bmatrix} = (k_1 + k_2) \cdot \begin{bmatrix} 1 & -1 \\ 2 & 3 \end{bmatrix} \in W$

   3. Let $c \in \mathbb{R}$ and $k \cdot \begin{bmatrix} 1 & -1 \\ 2 & 3 \end{bmatrix} \in W$.
      Then $c \cdot \left(k \cdot \begin{bmatrix} 1 & -1 \\ 2 & 3 \end{bmatrix} \right) = (ck) \cdot \begin{bmatrix} 1 & -1 \\ 2 & 3 \end{bmatrix} \in W$

   Now, to show that W is nontrivial, we need to show that W contains more than just the zero matrix and that W is not all of $M_{2,2}$.

   Let $k = 1$. Then $k \cdot \begin{bmatrix} 1 & -1 \\ 2 & 3 \end{bmatrix} = \begin{bmatrix} 1 & -1 \\ 2 & 3 \end{bmatrix} \in W$

   $\Rightarrow W \neq \left\{ \begin{bmatrix} 0 & 0 \\ 0 & 0 \end{bmatrix} \right\}$.

   $\begin{bmatrix} 2 & -2 \\ 4 & 9 \end{bmatrix} \in M_{2,2}$, but $\begin{bmatrix} 2 & -2 \\ 4 & 9 \end{bmatrix}$ does not lie in W.

   $\Rightarrow W \neq M_{2,2}$.

SUBSETS THAT ARE NOT SUBSPACES

Now we're going to look at subsets that are not subspaces. Let's take a look at an example.

Let $W = \{(x, x^2) | x \in \mathbb{R}\}$. Notice that W is a subset of $\mathbb{R}^2$. We can write it like this $W \subseteq \mathbb{R}^2$. Notice that the second coordinate is x^2; so we have a parabola $y = x^2$ in the plane. W is the set of all points on the parabola. It is a subset of $\mathbb{R}^2$ because it's contained in $\mathbb{R}^2$; but, it's not a subspace of $\mathbb{R}^2$. We'll see this by checking the three subspace properties; if any of the properties fail, then we know it's not a subspace.

1. Let $x = 0$. Then $(x, x^2) = (0,0)$. So $\mathbf{0} \in W$.
2. Let $u = (x, x^2)$ and $v = (y, y^2)$. Then $u + v = (x + y, x^2 + y^2)$. The question is, does this lie in W? It would if the second coordinate is the square of the first component. However, usually $x^2 + y^2 \neq (x + y)^2$. For instance, $(1, 1)$ and $(2, 4)$ lie in W, but $(1,1) + (2,4) = (3,5)$, which doesn't lie in W because $3^2 = 9 \neq 5$. So closure under addition fails. We can stop right here and conclude that W is not a subspace of $\mathbb{R}^2$. Let's check the third subspace property anyways.
3. Let $u = (x, x^2) \in W$ and $c \in \mathbb{R}$.
   Then $cu = (cx, cx^2)$. The question is, does this lie in W? It would if $cx^2 = (cx)^2$. However, usually $cx^2 \neq (cx)^2$. For instance, let $c = 3$ and $u = (2, 4)$. Then $cu = (6,12)$, and $12 \neq 6^2$. So closure under scalar multiplication fails.

Since at least one of the subspace properties fails, W is not a subspace of $\mathbb{R}^2$.

SUBSETS THAT ARE NOT SUBSPACES: ADDITIONAL EXAMPLE

Let's look at another example of a subset that's not a subspace.

Let W be the set of all 2×2 matrices that are not invertible. Show that W is not a subspace of $M_{2,2}$. Let's check the three subspace properties.

1. The zero matrix lies in W since $\begin{bmatrix} 0 & 0 \\ 0 & 0 \end{bmatrix}$ is not invertible. The determinant of $\begin{bmatrix} 0 & 0 \\ 0 & 0 \end{bmatrix}$ is 0.
2. Let $A = \begin{bmatrix} 2 & 6 \\ 1 & 3 \end{bmatrix}$ and $B = \begin{bmatrix} -1 & -6 \\ -1 & -6 \end{bmatrix}$. Then A and B are both not invertible since $\det(A) = 0$ and $\det(B) = 0$. So A and B lie in W.
   Now, $A + B = \begin{bmatrix} 1 & 0 \\ 0 & -3 \end{bmatrix}$ and $\det(A + B) = -3 \neq 0$. So $A + B$ is invertible. Therefore, $A + B$ does not lie in W. We can conclude that closure under addition fails.

Since at least one of the subspace properties fails, W is not a subspace of $M_{2,2}$.

PROBLEM SET: SUBSETS THAT ARE NOT SUBSPACES

1. Show that the set $W = \{(x,y) | x, y \in \mathbb{R}, y \geq 0\}$, the upper half of the plane, is not a subspace of $\mathbb{R}^2$.

2. Show that the set $W = \{(x,y) | x, y \in \mathbb{R} \text{ and } x^2 + y^2 = 1\}$, the circle of radius 1 in the plane, is not a subspace of $\mathbb{R}^2$.

3. Show that the set W of all n by n matrices with determinant zero is not a subspace of $M_{n,n}$.

SOLUTION SET: SUBSETS THAT ARE NOT SUBSPACES

1. $W = \{(x,y) | x, y \in \mathbb{R}, y \geq 0\}$
   Closure under scalar multiplication fails.

   $(1,1) \in W$. Let $c = -1$. Then $c(1,1) = (-1)(1,1) = (-1,-1)$, which does not lie in W.

2. $W = \{(x,y) | x, y \in \mathbb{R} \text{ and } x^2 + y^2 = 1\}$
   $(0,0)$ does not lie in W. Also, closure under scalar multiplication fails. $(1,0) \in W$, but $\frac{1}{2}(1,0) = (\frac{1}{2}, 0)$, which does not lie in W.

3. Let $A = \begin{bmatrix} 1 & 1 \\ 2 & 2 \end{bmatrix}$ and $B = \begin{bmatrix} -1 & 0 \\ 6 & 0 \end{bmatrix}$. Then, $\det(A) = 0$ and $\det(B) = 0$. So $A, B \in W$.
   $A + B = \begin{bmatrix} 0 & 1 \\ 8 & 2 \end{bmatrix}$. $\det(A + B) = -8 \neq 0$. So $A + B$ does not lie in W. Closure under addition fails.

SUMMARY: SUBSPACES

- A subset W of a vector space V is a **subspace** of V if W is nonempty and a vector space itself with the same operations as V.

- The subspace properties consist of:
    4. $\mathbf{0} \in W$
    5. $u + v \in W$ for all $u, v \in W$.   Closure under addition
    6. $cu \in W$ for all $c \in \mathbb{R}, u \in W$.   Closure under scalar multiplication

- The subset consisting of only the zero vector $\{\mathbf{0}\}$ is a subspace of V. It's called the **zero subspace**. The set V itself is a subspace of V. These two subspaces are called **trivial subspaces** of V.

- A **nontrivial subspace** is any subspace that is not the zero subspace or the whole vector space V.

- If a subset W of a vector space V fails to satisfy any of the three subspace properties, then W is not a subspace of V.

11 – SPAN AND LINEAR INDEPENDENCE

SPAN

In this section, we're going to explore the notions of span and linear independence. First, let's look at the notion of span.

Let V be a vector space and $S = \{v_1, v_2, \ldots, v_k\}$ a subset of V. If every vector in V can be written as a linear combination of vectors in S, then we say that S *spans* V.

For example, show that $S = \{(1,0), (0,1)\}$ spans $\mathbb{R}^2$.

Let $(u_1, u_2) \in \mathbb{R}^2$, where $u_1, u_2 \in \mathbb{R}$.

Then we can rewrite (u_1, u_2) as $u_1(1,0) + u_2(0,1)$, which is a linear combination of the vectors $(1,0)$ and $(0,1)$. So S spans $\mathbb{R}^2$.

Let's do another example.

Let $S = \left\{ \begin{bmatrix} 1 & 0 \\ 0 & 0 \end{bmatrix}, \begin{bmatrix} 0 & 1 \\ 0 & 0 \end{bmatrix}, \begin{bmatrix} 0 & 0 \\ 1 & 0 \end{bmatrix}, \begin{bmatrix} 0 & 0 \\ 0 & 1 \end{bmatrix} \right\}$. Show that S spans $M_{2,2}$, the set of all 2×2 matrices.

Let $\begin{bmatrix} a & b \\ c & d \end{bmatrix} \in M_{2,2}$, where $a, b, c, d \in \mathbb{R}$.

Then we can rewrite $\begin{bmatrix} a & b \\ c & d \end{bmatrix}$ as $a\begin{bmatrix} 1 & 0 \\ 0 & 0 \end{bmatrix} + b\begin{bmatrix} 0 & 1 \\ 0 & 0 \end{bmatrix} + c\begin{bmatrix} 0 & 0 \\ 1 & 0 \end{bmatrix} + d\begin{bmatrix} 0 & 0 \\ 0 & 1 \end{bmatrix}$, which is a linear combination of the vectors in S. So S spans $M_{2,2}$.

Now, let's look at some examples of subsets that do not span the entire space.

Show that $S = \{(1,0), (3,0)\}$ does not span $\mathbb{R}^2$.

To show this, we just need to find one vector in $\mathbb{R}^2$ that cannot be written as a linear combination of vectors in S.

Consider $(2,2)$. If we could write $(2,2)$ as a linear combination of $(1,0)$ and $(3,0)$, we would have $c_1(1,0) + c_2(3,0) = (2,2)$ for some scalars c_1 and c_2.

$\Rightarrow (c_1, 0) + (3c_2, 0) = (2,2)$

$\Rightarrow (c_1 + 3c_2, 0) = (2,2)$

$\Rightarrow c_1 + 3c_2 = 2$ and $0 = 2$. The second equation is a contradiction. So we cannot write $(2,2)$ as a linear combination of $(1,0)$ and $(3,0)$. Since we found one vector, $(2,2)$, in $\mathbb{R}^2$ that cannot be written as a linear combination of the vectors in S, S does not span $\mathbb{R}^2$.

Let's do another example.

Suppose $S = \left\{ \begin{bmatrix} 1 & 1 \\ 0 & 0 \end{bmatrix}, \begin{bmatrix} 1 & -1 \\ 2 & 0 \end{bmatrix}, \begin{bmatrix} 0 & 0 \\ 1 & 1 \end{bmatrix}, \begin{bmatrix} 2 & -3 \\ 0 & -5 \end{bmatrix} \right\}$. Show that S does not span $M_{2,2}$.

Consider $\begin{bmatrix} 1 & 2 \\ 4 & 1 \end{bmatrix}$. If this matrix could be written as a linear combination of the matrices in S, then we would have $c_1 \begin{bmatrix} 1 & 1 \\ 0 & 0 \end{bmatrix} + c_2 \begin{bmatrix} 1 & -1 \\ 2 & 0 \end{bmatrix} + c_3 \begin{bmatrix} 0 & 0 \\ 1 & 1 \end{bmatrix} + c_4 \begin{bmatrix} 2 & -3 \\ 0 & -5 \end{bmatrix} = \begin{bmatrix} 1 & 2 \\ 4 & 1 \end{bmatrix}$, for some scalars c_1, c_2, c_3, c_4.

$$\implies \begin{bmatrix} c_1 & c_1 \\ 0 & 0 \end{bmatrix} + \begin{bmatrix} c_2 & -c_2 \\ 2c_2 & 0 \end{bmatrix} + \begin{bmatrix} 0 & 0 \\ c_3 & c_3 \end{bmatrix} + \begin{bmatrix} 2c_4 & -3c_4 \\ 0 & -5c_4 \end{bmatrix} = \begin{bmatrix} 1 & 2 \\ 4 & 1 \end{bmatrix}$$

$$\implies \begin{bmatrix} c_1 + c_2 + 2c_4 & c_1 - c_2 - 3c_4 \\ 2c_2 + c_3 & c_3 - 5c_4 \end{bmatrix} = \begin{bmatrix} 1 & 2 \\ 4 & 1 \end{bmatrix}$$

$\implies c_1 + c_2 + 2c_4 = 1$

$c_1 - c_2 - 3c_4 = 2$

$2c_2 + c_3 = 4$

$c_3 - 5c_4 = 1$

Form the augmented matrix for this system of equations.

$$\begin{bmatrix} 1 & 1 & 0 & 2 & 1 \\ 1 & -1 & 0 & -3 & 2 \\ 0 & 2 & 1 & 0 & 4 \\ 0 & 0 & 1 & -5 & 1 \end{bmatrix}$$

Try solving this by Gaussian elimination.

Perform $R1 - R2 \rightarrow R2$:

$$\begin{bmatrix} 1 & 1 & 0 & 2 & 1 \\ 0 & 2 & 0 & 5 & -1 \\ 0 & 2 & 1 & 0 & 4 \\ 0 & 0 & 1 & -5 & 1 \end{bmatrix}$$

Perform $R2 - R3 \rightarrow R3$:

$$\begin{bmatrix} 1 & 1 & 0 & 2 & 1 \\ 0 & 2 & 0 & 5 & -1 \\ 0 & 0 & -1 & 5 & -5 \\ 0 & 0 & 1 & -5 & 1 \end{bmatrix}$$

Perform $R3 + R4 \rightarrow R4$:

$$\begin{bmatrix} 1 & 1 & 0 & 2 & 1 \\ 0 & 2 & 0 & 5 & -1 \\ 0 & 0 & -1 & 5 & -5 \\ 0 & 0 & 0 & 0 & -4 \end{bmatrix}$$

In the last row, we get $0 = -4$, which is a contradiction. So the system has no solution, and $\begin{bmatrix} 1 & 2 \\ 4 & 1 \end{bmatrix}$ cannot be written as a linear combination of the matrices in S. So S does not span $M_{2,2}$.

SPAN OF A SUBSET OF A VECTOR SPACE

We've seen some examples of subsets S of a vector space V that do not span all of V. However, if we take the set of all linear combinations of vectors in $S = \{v_1, \ldots, v_k\}$, this set will form a subspace of V. The set of all linear combinations of vectors in S is called the ***span of S*** and is denoted $span(S)$.

Recall our earlier example $S = \{(1,0), (3,0)\}$ and $V = \mathbb{R}^2$. We saw that S does not span all of $\mathbb{R}^2$. However, if we take the set of all linear combinations of $(1,0)$ and $(3,0)$, we get the $span(S)$.

$$span(S) = \{c_1(1,0) + c_2(3,0) | c_1, c_2 \in \mathbb{R}\}$$

$$= \{k(1,0) | k \in \mathbb{R}\} \quad \text{(we can show this below)}$$

If $c_1(1,0) + c_2(3,0) \in span(S)$, then $c_1(1,0) + c_2(3,0) = (c_1, 0) + (3c_2, 0)$

$$= (c_1 + 3c_2, 0)$$

$$= ((c_1 + 3c_2) \cdot 1, 0)$$

$$= (c_1 + 3c_2)(1,0)$$

$$= k(1,0) \text{ if we let } k = c_1 + 3c_2$$

$$\in \{k(1,0) | k \in \mathbb{R}\}$$

So $span(S) \subseteq \{k(1,0) | k \in \mathbb{R}\}$.

Let $k(1,0) \in \{k(1,0) | k \in \mathbb{R}\}$. Then $k(1,0) = k(1,0) + 0(3,0)$. Let $c_1 = k$ and $c_2 = 0$.

Then $k(1,0) = c_1(1,0) + c_2(3,0) \in span(S)$.

$\Rightarrow \{k(1,0) | k \in \mathbb{R}\} \subseteq span(S)$.

Since $span(S) \subseteq \{k(1,0) | k \in \mathbb{R}\}$ and $\{k(1,0) | k \in \mathbb{R}\} \subseteq span(S)$, we have $span(S) = \{k(1,0) | k \in \mathbb{R}\}$.

So the $span(S)$ consists of all scalar multiples of the point $(1,0)$ in the plane $\mathbb{R}^2$.

We can think of $(1,0)$ as an arrow in the plane going from $(0,0)$ to $(1,0)$.

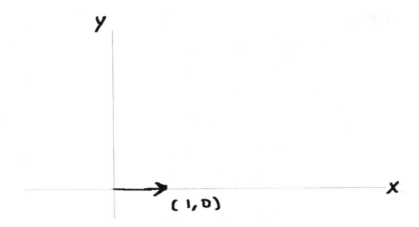

Scalar multiples of the vector (1,0) just give us more points on the real line. So $span(S)$ is the entire real line in the plane. Even though S doesn't span all of $\mathbb{R}^2$, it does span the entire real line and the span of S is a subspace of $\mathbb{R}^2$.

LINEAR INDEPENDENCE

The notion of linear independence, in addition to the notion of span, is an important notion in linear algebra.

Suppose $S = \{v_1, \ldots, v_k\}$ is a subset of a vector space V.

If the vector equation $c_1 v_1 + \cdots + c_k v_k = \mathbf{0}$ has only the trivial solution $c_1 = c_2 = \cdots = c_k = 0$, then the set S is said to be **linearly independent**. Otherwise, the set is said to be **linearly dependent**.

For example, let $S = \{(1,0), (0,1)\}$, a subset of $\mathbb{R}^2$. Show that S is linearly independent.

Suppose $c_1(1,0) + c_2(0,1) = \mathbf{0}$. Show that $c_1 = c_2 = 0$.

We have $c_1(1,0) + c_2(0,1) = (0,0)$

$\implies (c_1, 0) + (0, c_2) = (0,0)$

$\implies (c_1, c_2) = (0,0)$

$\implies c_1 = 0$ and $c_2 = 0$.

So the vector equation $c_1(1,0) + c_2(0,1) = \mathbf{0}$ has only the trivial solution. Therefore, S is linearly independent.

Let's do another example.

Suppose $S = \{(1,0), (3,0)\}$, a subset of $\mathbb{R}^2$. Show that S is linearly dependent.

Suppose $c_1(1,0) + c_2(3,0) = \mathbf{0}$. Find a non-trivial solution.

We have $(c_1, 0) + (3c_2, 0) = (0,0)$

$\implies (c_1 + 3c_2, 0) = (0,0)$

$\implies c_1 + 3c_2 = 0$ and $0 = 0$.

$\implies c_1 + 3c_2 = 0$

$\implies c_1 = -3c_2$. Let $c_2 = t$, a parameter.

$\implies c_1 - 3t$ and $c_2 = t$.

Let $t = 1$. Then $c_1 = -3$ and $c_2 = 1$. This is a nontrivial solution to the vector equation $c_1(1,0) + c_2(3,0) = \mathbf{0}$. You can check $-3(1,0) + 1(3,0) = \mathbf{0}$.

DETERMINING LINEAR INDEPENDENCE OR DEPENDENCE

Let's do some more examples.

Let $S = \{(1,-1,0), (2,1,1), (3,3,2)\}$, a subset of $\mathbb{R}^3$. Determine if S is linearly independent or linearly dependent.

Suppose $c_1(1,-1,0) + c_2(2,1,1) + c_3(3,3,2) = \mathbf{0}$.

Then $c_1 + 2c_2 + 3c_3 = 0$

$-c_1 + c_2 + 3c_3 = 0$

$c_2 + 2c_3 = 0$

So we have a system of equations. Let's form the augmented matrix and start doing Gaussian elimination.

$$\begin{bmatrix} 1 & 2 & 3 & 0 \\ -1 & 1 & 3 & 0 \\ 0 & 1 & 2 & 0 \end{bmatrix}$$

Perform $R1 + R2 \to R2$:

$$\begin{bmatrix} 1 & 2 & 3 & 0 \\ 0 & 3 & 6 & 0 \\ 0 & 1 & 2 & 0 \end{bmatrix}$$

Perform $\frac{1}{3}R2 \to R2$:

$$\begin{bmatrix} 1 & 2 & 3 & 0 \\ 0 & 1 & 2 & 0 \\ 0 & 1 & 2 & 0 \end{bmatrix}$$

Perform $R2 - R3 \rightarrow R3$:

$$\begin{bmatrix} 1 & 2 & 3 & 0 \\ 0 & 1 & 2 & 0 \\ 0 & 0 & 0 & 0 \end{bmatrix}$$

From the second row, $c_2 + 2c_3 = 0$. Let $c_3 = t$. Then $c_2 = -2t$.

From the first row, $c_1 + 2c_2 + 3c_3 = 0$.

$\Rightarrow c_1 + 2(-2t) + 3t = 0$

$\Rightarrow c_1 - 4t + 3t = 0$

$\Rightarrow c_1 - t = 0$

$\Rightarrow c_1 = t$

$\Rightarrow c_1 = t, c_2 = -2t, c_3 = t$.

Let $t = 1$.

$\Rightarrow c_1 = 1, c_2 = -2, c_3 = 1$. This is a nontrivial solution to the original vector equation. So S is linearly dependent.

Let's do another example.

Let $S = \{1 + 2x + 3x^2, x + 2x^2, -2 + x^2\}$, a subset of P_2. P_2 is the set of all polynomials of degree less than or equal to 2.

Determine if S is linearly independent or linearly dependent.

Suppose $c_1(1 + 2x + 3x^2) + c_2(x + 2x^2) + c_3(-2 + x^2) = \mathbf{0} = 0 + 0x + 0x^2$.

$\Rightarrow (c_1 + 2c_1x + 3c_1x^2) + (c_2x + 2c_2x^2) + (-2c_3 + c_3x^2) = 0 + 0x + 0x^2$

$\Rightarrow (c_1 - 2c_3) + (2c_1 + c_2)x + (3c_1 + 2c_2 + c_3)x^2 = 0 + 0x + 0x^2$

$\Rightarrow c_1 - 2c_3 = 0$

$2c_1 + c_2 = 0$

$3c_1 + 2c_2 + c_3 = 0$

Form the augmented matrix and perform Gaussian elimination.

$$\begin{bmatrix} 1 & 0 & -2 & 0 \\ 2 & 1 & 0 & 0 \\ 3 & 2 & 1 & 0 \end{bmatrix}$$

Perform $-2R1 + R2 \rightarrow R2$:

$$\begin{bmatrix} 1 & 0 & -2 & 0 \\ 0 & 1 & 4 & 0 \\ 3 & 2 & 1 & 0 \end{bmatrix}$$

Perform $-3R1 + R3 \rightarrow R3$:

$$\begin{bmatrix} 1 & 0 & -2 & 0 \\ 0 & 1 & 4 & 0 \\ 0 & 2 & 7 & 0 \end{bmatrix}$$

Perform $-2R2 + R3 \rightarrow R3$:

$$\begin{bmatrix} 1 & 0 & -2 & 0 \\ 0 & 1 & 4 & 0 \\ 0 & 0 & -1 & 0 \end{bmatrix}$$

Perform $-R3 \rightarrow R3$:

$$\begin{bmatrix} 1 & 0 & -2 & 0 \\ 0 & 1 & 4 & 0 \\ 0 & 0 & 1 & 0 \end{bmatrix}$$

Row 3 entails that $c_3 = 0$.

Row 2 entails that $c_2 + 4c_3 = 0. \implies c_2 + 4(0) = 0$

$$\implies c_2 = 0$$

Row 1 entails that $c_1 - 2c_3 = 0$

$\implies c_1 - 2(0) = 0$

$\implies c_1 = 0$

So $c_1 = c_2 = c_3 = 0$. The system has only the trivial solution. So S is linearly independent.

PROBLEM SET: SPAN AND LINEAR INDEPENDENCE

1. Show that $S = \left\{ \begin{bmatrix} 1 & 0 & 0 \\ 0 & 0 & 0 \end{bmatrix}, \begin{bmatrix} 0 & 1 & 0 \\ 0 & 0 & 0 \end{bmatrix}, \begin{bmatrix} 0 & 0 & 1 \\ 0 & 0 & 0 \end{bmatrix}, \begin{bmatrix} 0 & 0 & 0 \\ 1 & 0 & 0 \end{bmatrix}, \begin{bmatrix} 0 & 0 & 0 \\ 0 & 1 & 0 \end{bmatrix}, \begin{bmatrix} 0 & 0 & 0 \\ 0 & 0 & 1 \end{bmatrix} \right\}$ spans $M_{2,3}$, where $M_{2,3}$ is the set of all 2 by 3 matrices.

2. Show that $S = \{1, x, 1 + x\}$ does not span P_2, where P_2 is the set of all polynomials of degree less than or equal to 2.

3. Let $S = \{1, x - x^2, x + x^2\}$, a subset of P_2. Show that S is linearly independent.

4. Let $S = \left\{ \begin{bmatrix} 0 & 0 \\ 1 & 1 \end{bmatrix}, \begin{bmatrix} 1 & 0 \\ 0 & 1 \end{bmatrix}, \begin{bmatrix} -1 & 0 \\ 0 & 0 \end{bmatrix}, \begin{bmatrix} -5 & 0 \\ 2 & -1 \end{bmatrix} \right\}$, a subset of $M_{2,2}$. Show that S is linearly dependent.

5. Let $S = \left\{ \begin{bmatrix} 1 & 0 \\ 1 & 0 \end{bmatrix}, \begin{bmatrix} 0 & 1 \\ 0 & 1 \end{bmatrix}, \begin{bmatrix} 1 & 1 \\ 0 & 0 \end{bmatrix}, \begin{bmatrix} 0 & 0 \\ 1 & 1 \end{bmatrix} \right\}$, a subset of $M_{2,2}$. Determine if S is linearly independent or linearly dependent.

SOLUTION SET: SPAN AND LINEAR INDEPENDENCE

1. Let $\begin{bmatrix} a & b & c \\ d & e & f \end{bmatrix} \in M_{2,3}$, where $a, b, c, d, e, f \in \mathbb{R}$.

   $\Rightarrow \begin{bmatrix} a & b & c \\ d & e & f \end{bmatrix} =$

   $a\begin{bmatrix} 1 & 0 & 0 \\ 0 & 0 & 0 \end{bmatrix} + b\begin{bmatrix} 0 & 1 & 0 \\ 0 & 0 & 0 \end{bmatrix} + c\begin{bmatrix} 0 & 0 & 1 \\ 0 & 0 & 0 \end{bmatrix} + d\begin{bmatrix} 0 & 0 & 0 \\ 1 & 0 & 0 \end{bmatrix} + e\begin{bmatrix} 0 & 0 & 0 \\ 0 & 1 & 0 \end{bmatrix} + f\begin{bmatrix} 0 & 0 & 0 \\ 0 & 0 & 1 \end{bmatrix}$

   $\Rightarrow$ S spans $M_{2,3}$

2. We need to find one polynomial in P_2 that cannot be written as a linear combination of polynomials in S.

   Consider $p(x) = x^2$. Suppose $x^2 = c_1 \cdot 1 + c_2 \cdot x + c_3 \cdot (1 + x)$ for some scalars c_1, c_2, c_3.

   Then $x^2 = (c_1 + c_3) + (c_2 + c_3)x + 0 \cdot x^2$.

   $\Rightarrow 0 \cdot 1 + 0 \cdot x + 1 \cdot x^2 = (c_1 + c_3) + (c_2 + c_3)x + 0 \cdot x^2$

   $\Rightarrow 0 = c_1 + c_3$

   $0 = c_2 + c_3$

   $1 = 0$

   $\Rightarrow$ Contradiction.

   $\Rightarrow$ S does not span P_2.

3. Suppose $c_1 \cdot 1 + c_2(x - x^2) + c_3(x + x^2) = 0$. Show that $c_1 = c_2 = c_3 = 0$.

   $c_1 \cdot 1 + c_2(x - x^2) + c_3(x + x^2) = 0$.

   $\Rightarrow c_1 + (c_2 + c_3)x + (-c_2 + c_3)x^2 = 0$

   $\Rightarrow c_1 = 0$

   $c_2 + c_3 = 0$

   $-c_2 + c_3 = 0$

   Solving this system of equations gives $c_1 = c_2 = c_3 = 0$.

4. Suppose $c_1\begin{bmatrix} 0 & 0 \\ 1 & 1 \end{bmatrix} + c_2\begin{bmatrix} 1 & 0 \\ 0 & 1 \end{bmatrix} + c_3\begin{bmatrix} -1 & 0 \\ 0 & 0 \end{bmatrix} + c_4\begin{bmatrix} -5 & 0 \\ 2 & -1 \end{bmatrix} = \begin{bmatrix} 0 & 0 \\ 0 & 0 \end{bmatrix}$.

   Find a non-trivial solution.

   $c_1\begin{bmatrix} 0 & 0 \\ 1 & 1 \end{bmatrix} + c_2\begin{bmatrix} 1 & 0 \\ 0 & 1 \end{bmatrix} + c_3\begin{bmatrix} -1 & 0 \\ 0 & 0 \end{bmatrix} + c_4\begin{bmatrix} -5 & 0 \\ 2 & -1 \end{bmatrix} = \begin{bmatrix} 0 & 0 \\ 0 & 0 \end{bmatrix}$

$$\Rightarrow \begin{bmatrix} 0 & 0 \\ c_1 & c_1 \end{bmatrix} + \begin{bmatrix} c_2 & 0 \\ 0 & c_2 \end{bmatrix} + \begin{bmatrix} -c_3 & 0 \\ 0 & 0 \end{bmatrix} + \begin{bmatrix} -5c_4 & 0 \\ 2c_4 & -c_4 \end{bmatrix} = \begin{bmatrix} 0 & 0 \\ 0 & 0 \end{bmatrix}$$

$$\Rightarrow \begin{bmatrix} c_2 - c_3 - 5c_4 & 0 \\ c_1 + 2c_4 & c_1 + c_2 - c_4 \end{bmatrix} = \begin{bmatrix} 0 & 0 \\ 0 & 0 \end{bmatrix}$$

$$\Rightarrow c_2 - c_3 - 5c_4 = 0$$
$$0 = 0$$
$$c_1 + 2c_4 = 0$$
$$c_1 + c_2 - c_4 = 0$$

$$\Rightarrow \begin{bmatrix} 0 & 1 & -1 & -5 \\ 0 & 0 & 0 & 0 \\ 1 & 0 & 0 & 2 \\ 1 & 1 & 0 & -1 \end{bmatrix} \begin{bmatrix} c_1 \\ c_2 \\ c_3 \\ c_4 \end{bmatrix} = \begin{bmatrix} 0 \\ 0 \\ 0 \\ 0 \end{bmatrix}$$

Perform Gauss-Jordan elimination on $\begin{bmatrix} 0 & 1 & -1 & -5 \\ 0 & 0 & 0 & 0 \\ 1 & 0 & 0 & 2 \\ 1 & 1 & 0 & -1 \end{bmatrix}$ to get $\begin{bmatrix} 1 & 0 & 0 & 2 \\ 0 & 1 & 0 & -3 \\ 0 & 0 & 1 & 2 \\ 0 & 0 & 0 & 0 \end{bmatrix}$.

The third row gives $c_3 + 2c_4 = 0$. Let $c_4 = t$. Then $c_3 = -2t$. The second row gives $c_2 - 3c_4 = 0$. So $c_2 = 3t$. The first row gives $c_1 + 2c_4 = 0$. So $c_1 = -2t$.

$$\Rightarrow \begin{bmatrix} c_1 \\ c_2 \\ c_3 \\ c_4 \end{bmatrix} = \begin{bmatrix} -2t \\ 3t \\ -2t \\ t \end{bmatrix} = t \begin{bmatrix} -2 \\ 3 \\ -2 \\ 1 \end{bmatrix}$$

Let $t = 1$. Then $\begin{bmatrix} c_1 \\ c_2 \\ c_3 \\ c_4 \end{bmatrix} = \begin{bmatrix} -2 \\ 3 \\ -2 \\ 1 \end{bmatrix}$ is a nontrivial solution to the equation

$$c_1 \begin{bmatrix} 0 & 0 \\ 1 & 1 \end{bmatrix} + c_2 \begin{bmatrix} 1 & 0 \\ 0 & 1 \end{bmatrix} + c_3 \begin{bmatrix} -1 & 0 \\ 0 & 0 \end{bmatrix} + c_4 \begin{bmatrix} -5 & 0 \\ 2 & -1 \end{bmatrix} = \begin{bmatrix} 0 & 0 \\ 0 & 0 \end{bmatrix}$$

$\Rightarrow$ S is linearly dependent.

5. Suppose $c_1 \begin{bmatrix} 1 & 0 \\ 1 & 0 \end{bmatrix} + c_2 \begin{bmatrix} 0 & 1 \\ 0 & 1 \end{bmatrix} + c_3 \begin{bmatrix} 1 & 1 \\ 0 & 0 \end{bmatrix} + c_4 \begin{bmatrix} 0 & 0 \\ 1 & 1 \end{bmatrix} = \begin{bmatrix} 0 & 0 \\ 0 & 0 \end{bmatrix}$.

$$\Rightarrow \begin{bmatrix} c_1 + c_3 & c_2 + c_3 \\ c_1 + c_4 & c_2 + c_4 \end{bmatrix} = \begin{bmatrix} 0 & 0 \\ 0 & 0 \end{bmatrix}$$

$$\Rightarrow c_1 + c_3 = 0$$

$$c_2 + c_3 = 0$$

$$c_1 + c_4 = 0$$

$$c_2 + c_4 = 0$$

$$\Rightarrow \begin{bmatrix} 1 & 0 & 1 & 0 \\ 0 & 1 & 1 & 0 \\ 1 & 0 & 0 & 1 \\ 0 & 1 & 0 & 1 \end{bmatrix} \begin{bmatrix} c_1 \\ c_2 \\ c_3 \\ c_4 \end{bmatrix} = \begin{bmatrix} 0 \\ 0 \\ 0 \\ 0 \end{bmatrix}$$

Perform Gauss-Jordan elimination on $\begin{bmatrix} 1 & 0 & 1 & 0 \\ 0 & 1 & 1 & 0 \\ 1 & 0 & 0 & 1 \\ 0 & 1 & 0 & 1 \end{bmatrix}$ to get $\begin{bmatrix} 1 & 0 & 0 & 1 \\ 0 & 1 & 0 & 1 \\ 0 & 0 & 1 & -1 \\ 0 & 0 & 0 & 0 \end{bmatrix}$.

Row 3 says $c_3 - c_4 = 0$. Let $c_4 = t$. So $c_3 = t$. Row 2 says $c_2 + c_4 = 0$. So $c_2 = -t$. Row 1 says $c_1 + c_4 = 0$. So $c_1 = -t$.

$$\Rightarrow \begin{bmatrix} c_1 \\ c_2 \\ c_3 \\ c_4 \end{bmatrix} = \begin{bmatrix} -t \\ -t \\ t \\ t \end{bmatrix} = t \begin{bmatrix} -1 \\ -1 \\ 1 \\ 1 \end{bmatrix}.$$

Let $t = 1$. Then $\begin{bmatrix} c_1 \\ c_2 \\ c_3 \\ c_4 \end{bmatrix} = \begin{bmatrix} -1 \\ -1 \\ 1 \\ 1 \end{bmatrix}$ is a nontrivial solution to the equation

$$c_1 \begin{bmatrix} 1 & 0 \\ 1 & 0 \end{bmatrix} + c_2 \begin{bmatrix} 0 & 1 \\ 0 & 1 \end{bmatrix} + c_3 \begin{bmatrix} 1 & 1 \\ 0 & 0 \end{bmatrix} + c_4 \begin{bmatrix} 0 & 0 \\ 1 & 1 \end{bmatrix} = \begin{bmatrix} 0 & 0 \\ 0 & 0 \end{bmatrix}$$

$\Rightarrow$ S is linearly dependent.

SUMMARY: SPAN AND LINEAR INDEPENDENCE

- Let V be a vector space and $S = \{v_1, v_2, \ldots, v_k\}$ a subset of V. If every vector in V can be written as a linear combination of vectors in S, then we say that S ***spans*** V.

- The set of all linear combinations of vectors in S is called the ***span of S*** and is denoted $span(S)$.

- If the vector equation $c_1 v_1 + \cdots + c_k v_k = \mathbf{0}$ has only the trivial solution $c_1 = c_2 = \cdots = c_k = 0$, then the set S is said to be ***linearly independent***. Otherwise, the set is said to be ***linearly dependent***.

12 – BASIS AND DIMENSION

BASIS

So far, we've seen that a subset $S = \{v_1, \ldots, v_k\}$ of a vector space V can span all of V. We've also seen what it means for S to be linearly independent. If S both spans V and is linearly independent, then S is said to be a ***basis*** for V.

For example, the set $S = \{(1,0), (0,1)\}$ is a basis for $\mathbb{R}^2$. We've seen, in earlier examples, that S spans $\mathbb{R}^2$ and is linearly independent. In fact, this basis is called the ***standard basis*** for $\mathbb{R}^2$. For $\mathbb{R}^3$, the standard basis is $\{(1,0,0), (0,1,0), (0,0,1)\}$. For $\mathbb{R}^n$, in general, the standard basis is $\{(1,0,\ldots,0), (0,1,0,\ldots,0), \ldots, (0,0,\ldots,1)\}$, where there are n vectors in the set.

A vector space could have a non-standard basis. For example, show that $S = \{(1,2), (-2,3)\}$ is a non-standard basis for $\mathbb{R}^2$.

To show that S is a basis, we need to show that S is linearly independent and spans $\mathbb{R}^2$. First, let's show that it's linearly independent.

Suppose $c_1(1,2) + c_2(-2,3) = (0,0)$.

$\Rightarrow (c_1 - 2c_2, 2c_1 + 3c_2) = \mathbf{0}$

$\Rightarrow c_1 - 2c_2 = 0$

$\quad 2c_1 + 3c_2 = 0$

Form the augmented matrix and perform Gaussian elimination:

$\begin{bmatrix} 1 & -2 & 0 \\ 2 & 3 & 0 \end{bmatrix}$

Perform $-2R1 + R2 \to R2$.

$\begin{bmatrix} 1 & -2 & 0 \\ 0 & 7 & 0 \end{bmatrix}$

$\Rightarrow c_2 = 0$

$\Rightarrow c_1 = 0$

Since the system has only the trivial solution, S is linearly independent.

Now, show that S spans $\mathbb{R}^2$.

Let $(u_1, u_2) \in \mathbb{R}^2$. We need to show that there are scalars c_1 and c_2 such that $(u_1, u_2) = c_1(1,2) + c_2(-2,3)$.

$c_1(1,2) + c_2(-2,3) = (c_1 - 2c_2, 2c_1 + 3c_2)$. So we want c_1 and c_2 such that $(c_1 - 2c_2, 2c_1 + 3c_2) = (u_1, u_2)$.

We get $c_1 - 2c_2 = u_1$

$2c_1 + 3c_2 = u_2$

We need a solution to this system $\begin{bmatrix} 1 & -2 & u_1 \\ 2 & 3 & u_2 \end{bmatrix}$. Perform $-2R1 + R2 \to R2$.

$\begin{bmatrix} 1 & -2 & u_1 \\ 0 & 7 & -2u_1 + u_2 \end{bmatrix}$

Perform $\frac{1}{7}R2 \to R2$:

$\begin{bmatrix} 1 & -2 & u_1 \\ 0 & 1 & (-2u_1 + u_2)/7 \end{bmatrix}$

Perform $2R2 + R1 \to R1$:

$\begin{bmatrix} 1 & 0 & \dfrac{3u_1 + 2u_2}{7} \\ 0 & 1 & \dfrac{-2u_1 + u_2}{7} \end{bmatrix}$

So $c_1 = \frac{3u_1 + 2u_2}{7}$ and $c_2 = \frac{-2u_1 + u_2}{7}$. We have a solution. Therefore, S spans $\mathbb{R}^2$.

Let's do another example. Show that $S = \left\{ \begin{bmatrix} 1 & 0 \\ 0 & 0 \end{bmatrix}, \begin{bmatrix} 0 & 1 \\ 0 & 0 \end{bmatrix}, \begin{bmatrix} 0 & 0 \\ 1 & 0 \end{bmatrix}, \begin{bmatrix} 0 & 0 \\ 0 & 1 \end{bmatrix} \right\}$ is a basis for $M_{2,2}$.

In a previous example, we've already shown that S spans $M_{2,2}$. We just need to show that S is linearly independent.

Suppose $c_1 \begin{bmatrix} 1 & 0 \\ 0 & 0 \end{bmatrix} + c_2 \begin{bmatrix} 0 & 1 \\ 0 & 0 \end{bmatrix} + c_3 \begin{bmatrix} 0 & 0 \\ 1 & 0 \end{bmatrix} + c_4 \begin{bmatrix} 0 & 0 \\ 0 & 1 \end{bmatrix} = \begin{bmatrix} 0 & 0 \\ 0 & 0 \end{bmatrix}$.

Then $\begin{bmatrix} c_1 & 0 \\ 0 & 0 \end{bmatrix} + \begin{bmatrix} 0 & c_2 \\ 0 & 0 \end{bmatrix} + \begin{bmatrix} 0 & 0 \\ c_3 & 0 \end{bmatrix} + \begin{bmatrix} 0 & 0 \\ 0 & c_4 \end{bmatrix} = \begin{bmatrix} 0 & 0 \\ 0 & 0 \end{bmatrix}$

$\Rightarrow \begin{bmatrix} c_1 & c_2 \\ c_3 & c_4 \end{bmatrix} = \begin{bmatrix} 0 & 0 \\ 0 & 0 \end{bmatrix}$

$\Rightarrow c_1 = c_2 = c_3 = c_4 = 0$.

So S is a basis for $M_{2,2}$. S is the standard basis for $M_{2,2}$.

The standard basis for P_n, the set of all polynomials of degree less than or equal to n, is $\{1, x, x^2, \ldots, x^n\}$.

As you can see, the vectors in a basis are like the building blocks for all other vectors in the vector space V. It turns out that, if $\{v_1, \ldots, v_k\}$ is a basis for V, not only can every vector in V be represented as a linear combination of $v_1, \ldots, v_k$, but the representation is unique. For instance, in P_2, the polynomial $3 + x - x^2$ can be written as a linear combination of the vectors $1, x, x^2$ in one and only one way, namely, as $3 \cdot 1 + 1 \cdot x + (-1) \cdot x^2$.

DIMENSION

One important fact about bases is that if a vector space V has a basis consisting of n vectors, then any other basis for V has n vectors. For example, we saw that $\{1, x, x^2\}$ is a basis for P_2. It turns out that $\{1 + 2x + 3x^2, x + 2x^2, -2 + x^2\}$ is also a basis for P_2. Note that it also has 3 vectors. Any other non-standard basis for P_2 will have 3 vectors.

We're now in a position to define the dimension of a vector space. If V has a basis consisting of n vectors, then the **dimension** of V is n. The dimension of V is unambiguous because every other basis for V has n vectors.

We've already seen that $\{(1,0), (0,1)\}$ is a basis for $\mathbb{R}^2$. So $\dim(\mathbb{R}^2) = 2$.

Consider the subspace $W = \{k(4,6) | k \in \mathbb{R}\}$. Let's find the dimension of W.

Every vector in W can be written as a scalar multiple of $(4,6)$. So the set $\{(4,6)\}$ spans W. The set $\{(4,6)\}$ is linearly independent. Let's show this.

If $c(4,6) = (0,0)$, then $(4c, 6c) = (0,0)$.

$\implies 4c = 0$ and $6c = 0$.

$\implies c = 0$.

So $\{(4,6)\}$ forms a basis for W. Therefore, $\dim W = 1$.

We can see this geometrically:

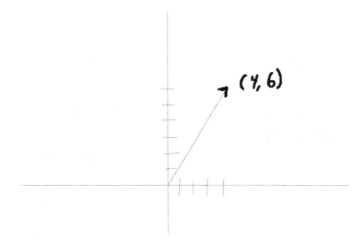

W consists of all scalar multiples of $(4, 6)$. So W consists of all points on the line going through the origin and $(4, 6)$.

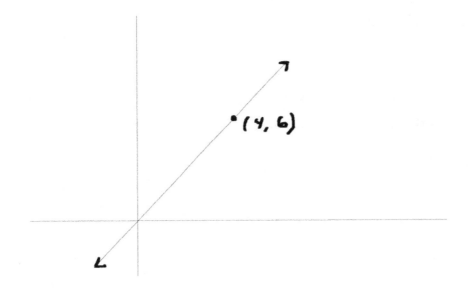

We can see that it makes sense that W is a 1-dimensional subspace of $\mathbb{R}^2$.

Let's do another example.

Let $W = \left\{ \begin{bmatrix} a & 0 & 0 \\ 0 & b & 0 \\ 0 & 0 & 0 \end{bmatrix} \mid a, b \in \mathbb{R} \right\}$. W is a subspace of $M_{3,3}$.

Find the dimension of W.

Note that
$$\begin{bmatrix} a & 0 & 0 \\ 0 & b & 0 \\ 0 & 0 & 0 \end{bmatrix} = \begin{bmatrix} a & 0 & 0 \\ 0 & 0 & 0 \\ 0 & 0 & 0 \end{bmatrix} + \begin{bmatrix} 0 & 0 & 0 \\ 0 & b & 0 \\ 0 & 0 & 0 \end{bmatrix}$$

$$= a \begin{bmatrix} 1 & 0 & 0 \\ 0 & 0 & 0 \\ 0 & 0 & 0 \end{bmatrix} + b \begin{bmatrix} 0 & 0 & 0 \\ 0 & 1 & 0 \\ 0 & 0 & 0 \end{bmatrix}$$

So every matrix in W can be written as a linear combination of $\begin{bmatrix} 1 & 0 & 0 \\ 0 & 0 & 0 \\ 0 & 0 & 0 \end{bmatrix}$ and $\begin{bmatrix} 0 & 0 & 0 \\ 0 & 1 & 0 \\ 0 & 0 & 0 \end{bmatrix}$. Thus, $\left\{ \begin{bmatrix} 1 & 0 & 0 \\ 0 & 0 & 0 \\ 0 & 0 & 0 \end{bmatrix}, \begin{bmatrix} 0 & 0 & 0 \\ 0 & 1 & 0 \\ 0 & 0 & 0 \end{bmatrix} \right\}$ spans W. It's easy to show that this set is also linearly independent.

Therefore, $\left\{ \begin{bmatrix} 1 & 0 & 0 \\ 0 & 0 & 0 \\ 0 & 0 & 0 \end{bmatrix}, \begin{bmatrix} 0 & 0 & 0 \\ 0 & 1 & 0 \\ 0 & 0 & 0 \end{bmatrix} \right\}$ forms a basis for W. Thus, $\dim W = 2$.

PROBLEM SET: BASIS AND DIMENSION

1. Show that $S = \left\{ \begin{bmatrix} 1 & 0 \\ 1 & 0 \end{bmatrix}, \begin{bmatrix} 1 & 0 \\ -1 & 0 \end{bmatrix}, \begin{bmatrix} 0 & 1 \\ 0 & 1 \end{bmatrix}, \begin{bmatrix} 0 & 1 \\ 0 & -1 \end{bmatrix} \right\}$ is a basis for $M_{2,2}$.

2. Let $W = \left\{ \begin{bmatrix} a & b \\ 0 & c \end{bmatrix} \mid a, b, c \in \mathbb{R} \right\}$. W is a subspace of $M_{2,2}$.
   Find the dimension of W.

SOLUTION SET: BASIS AND DIMENSION

1. First show that S is linearly independent.

   Suppose $c_1 \begin{bmatrix} 1 & 0 \\ 1 & 0 \end{bmatrix} + c_2 \begin{bmatrix} 1 & 0 \\ -1 & 0 \end{bmatrix} + c_3 \begin{bmatrix} 0 & 1 \\ 0 & 1 \end{bmatrix} + c_4 \begin{bmatrix} 0 & 1 \\ 0 & -1 \end{bmatrix} = \begin{bmatrix} 0 & 0 \\ 0 & 0 \end{bmatrix}$.

   $\Longrightarrow \begin{bmatrix} c_1 + c_2 & c_3 + c_4 \\ c_1 - c_2 & c_3 - c_4 \end{bmatrix} = \begin{bmatrix} 0 & 0 \\ 0 & 0 \end{bmatrix}$

   $\Longrightarrow$ $c_1 + c_2 = 0$
   $c_1 - c_2 = 0$
   $c_3 + c_4 = 0$
   $c_3 - c_4 = 0$

   $\Longrightarrow \begin{bmatrix} 1 & 1 & 0 & 0 \\ 1 & -1 & 0 & 0 \\ 0 & 0 & 1 & 1 \\ 0 & 0 & 1 & -1 \end{bmatrix} \begin{bmatrix} c_1 \\ c_2 \\ c_3 \\ c_4 \end{bmatrix} = \begin{bmatrix} 0 \\ 0 \\ 0 \\ 0 \end{bmatrix}$

   Perform Gauss-Jordan elimination on $\begin{bmatrix} 1 & 1 & 0 & 0 \\ 1 & -1 & 0 & 0 \\ 0 & 0 & 1 & 1 \\ 0 & 0 & 1 & -1 \end{bmatrix}$ to get $\begin{bmatrix} 1 & 0 & 0 & 0 \\ 0 & 1 & 0 & 0 \\ 0 & 0 & 1 & 0 \\ 0 & 0 & 0 & 1 \end{bmatrix}$.

   $\Longrightarrow$ $c_1 = c_2 = c_3 = c_4 = 0$.

   $\Longrightarrow$ S is linearly independent.

   Now, we need to show that S spans $M_{2,2}$.

   Let $\begin{bmatrix} a & b \\ c & d \end{bmatrix} \in M_{2,2}$. We need to show that there are scalars c_1, c_2, c_3, c_4 such that

   $\begin{bmatrix} a & b \\ c & d \end{bmatrix} = c_1 \begin{bmatrix} 1 & 0 \\ 1 & 0 \end{bmatrix} + c_2 \begin{bmatrix} 1 & 0 \\ -1 & 0 \end{bmatrix} + c_3 \begin{bmatrix} 0 & 1 \\ 0 & 1 \end{bmatrix} + c_4 \begin{bmatrix} 0 & 1 \\ 0 & -1 \end{bmatrix}$.

   $\Longrightarrow \begin{bmatrix} c_1 + c_2 & c_3 + c_4 \\ c_1 - c_2 & c_3 - c_4 \end{bmatrix} = \begin{bmatrix} a & b \\ c & d \end{bmatrix}$

   $\Longrightarrow$ $c_1 + c_2 = a$
   $c_1 - c_2 = c$
   $c_3 + c_4 = b$
   $c_3 - c_4 = d$

   We need a solution to this system of equations.

Form the augmented matrix $\begin{bmatrix} 1 & 1 & 0 & 0 & a \\ 1 & -1 & 0 & 0 & c \\ 0 & 0 & 1 & 1 & b \\ 0 & 0 & 1 & -1 & d \end{bmatrix}$.

Perform Gauss-Jordan elimination to get $\begin{bmatrix} 1 & 0 & 0 & 0 & \frac{a+c}{2} \\ 0 & 1 & 0 & 0 & \frac{a-c}{2} \\ 0 & 0 & 1 & 0 & \frac{b+d}{2} \\ 0 & 0 & 0 & 1 & \frac{b-d}{2} \end{bmatrix}$.

$\implies$ $c_1 = \frac{a+c}{2}$
$c_2 = \frac{a-c}{2}$
$c_3 = \frac{b+d}{2}$
$c_4 = \frac{b-d}{2}$

So we have a solution.

$\implies$ S spans $M_{2,2}$

$\implies$ S is a basis for $M_{2,2}$.

2. Note that $\begin{bmatrix} a & b \\ 0 & c \end{bmatrix} = a\begin{bmatrix} 1 & 0 \\ 0 & 0 \end{bmatrix} + b\begin{bmatrix} 0 & 1 \\ 0 & 0 \end{bmatrix} + c\begin{bmatrix} 0 & 0 \\ 0 & 1 \end{bmatrix}$.

So every matrix in W can be written as a linear combination of $\begin{bmatrix} 1 & 0 \\ 0 & 0 \end{bmatrix}, \begin{bmatrix} 0 & 1 \\ 0 & 0 \end{bmatrix}$, and $\begin{bmatrix} 0 & 0 \\ 0 & 1 \end{bmatrix}$.

Thus, $\left\{ \begin{bmatrix} 1 & 0 \\ 0 & 0 \end{bmatrix}, \begin{bmatrix} 0 & 1 \\ 0 & 0 \end{bmatrix}, \begin{bmatrix} 0 & 0 \\ 0 & 1 \end{bmatrix} \right\}$ spans W.

Now, let's show that this set is linearly independent.

Suppose $c_1 \begin{bmatrix} 1 & 0 \\ 0 & 0 \end{bmatrix} + c_2 \begin{bmatrix} 0 & 1 \\ 0 & 0 \end{bmatrix} + c_3 \begin{bmatrix} 0 & 0 \\ 0 & 1 \end{bmatrix} = \begin{bmatrix} 0 & 0 \\ 0 & 0 \end{bmatrix}$.

$\implies$ $\begin{bmatrix} c_1 & c_2 \\ 0 & c_3 \end{bmatrix} = \begin{bmatrix} 0 & 0 \\ 0 & 0 \end{bmatrix}$

$\implies$ $c_1 = c_2 = c_3 = 0$.

$\implies$ The set is linearly independent.

$\left\{ \begin{bmatrix} 1 & 0 \\ 0 & 0 \end{bmatrix}, \begin{bmatrix} 0 & 1 \\ 0 & 0 \end{bmatrix}, \begin{bmatrix} 0 & 0 \\ 0 & 1 \end{bmatrix} \right\}$ forms a basis for W.

$\implies$ $\dim W = 3$.

COORDINATES

If x is an arbitrary vector in V, and B is a basis for V, then x can be written as a linear combination of the vectors in B. So if $B = \{v_1, \ldots, v_n\}$, then $x = c_1 v_1 + \cdots + c_n v_n$ for some scalars $c_1, \ldots, c_n$. The scalars $c_1, \ldots, c_n$ are called the ***coordinates*** of x relative to the basis B. The vectors $v_1, \ldots, v_n$ in the basis B are like ingredients in a recipe, and the scalars $c_1, \ldots, c_n$ tell us the amount of each ingredient needed to cook up the vector x. We take the amount c_1 of v_1, the amount c_1 of v_2, etc. and add them all up to get $x = c_1 v_1 + c_2 v_2 + \cdots + c_n v_n$. For a different vector y in V, we're going to have different amounts for each ingredient to cook up y.

We can form a column matrix consisting of the coordinates of x relative to B as follows:

$$\begin{bmatrix} c_1 \\ c_2 \\ \vdots \\ c_n \end{bmatrix}$$

This is called the ***coordinate matrix*** of x relative to B and is denoted $[x]_B$.

CHANGE OF BASIS

We have seen that a vector space can have more than one basis. For example, $\{(1,0), (0,1)\}$ is a basis for $\mathbb{R}^2$, but so is $\{(1,2), (-2,3)\}$. The first one is the standard basis, and the second one is a non-standard basis. There could be many non-standard bases.

Let B be the standard basis $\{(1,0), (0,1)\}$ and let B' be the non-standard basis $\{(1,2), (-2,3)\}$. We want to be able to represent a vector in $\mathbb{R}^2$ given in terms of B as a vector in terms of B' and vice versa. In other words, we want to be able to change the basis.

For example, let $x = (4,15)$. $(4,15)$ can be rewritten as $4(1,0) + 15(0,1)$. So the coordinates for x relative to B are 4 and 15. So the coordinate matrix for x relative to B is $\begin{bmatrix} 4 \\ 15 \end{bmatrix}$.

We want to find the coordinates for x relative to B'.

We want $(4,15) = c_1(1,2) + c_2(-2,3)$.

$\Rightarrow (4,15) = (c_1 - 2c_2, 2c_1 + 3c_2)$

$\Rightarrow c_1 - 2c_2 = 4$

$ 2c_1 + 3c_2 = 15$

$\Rightarrow c_1 - 2c_2 = 4$

$ \quad 7c_2 = 7$ (after performing $-2R1 + R2 \to R2$)

$\Rightarrow c_2 = 1$ and $c_1 = 6$

$\Rightarrow (4,15) = 6(1,2) + 1(-2,3)$

⇒ The coordinates for x relative to B' are 6 and 1.

⇒ The coordinate matrix for x relative to B' is $\begin{bmatrix} 6 \\ 1 \end{bmatrix}$.

Thus, $[x]_B = \begin{bmatrix} 4 \\ 15 \end{bmatrix}$ and $[x]_{B'} = \begin{bmatrix} 6 \\ 1 \end{bmatrix}$.

In our example, we changed the basis from B to B' for the vector $x = (4,15)$. We want to be able to do this for any vector in $\mathbb{R}^2$. More generally, if V is an n-dimensional vector space and B and B' are two bases for V, we want to be able to change the basis from B to B' for any vector in V.

It turns out that there is a way to do this. There is a matrix P, called the **transition matrix** from B to B', such that $P[x]_B = [x]_{B'}$. If we're given the coordinate matrix for x relative to B, we can simply multiply by the transition matrix P and the result will be the coordinate matrix for x relative to B'.

There is a procedure for finding the transition matrix P. Form the augmented matrix $[B' \vdots B]$ and perform Gauss-Jordan elimination to get $[I_n \vdots P]$.

EXAMPLES OF FINDING TRANSITION MATRICES

Let's do some examples.

Find the transition matrix from B to B' where $B = \{(1,0), (0,1)\}$ and $B' = \{(1,2), (-2,3)\}$.

Form the augmented matrix $[B' \vdots B]$.

$$\begin{bmatrix} 1 & -2 & \vdots & 1 & 0 \\ 2 & 3 & \vdots & 0 & 1 \end{bmatrix}$$

Perform Gauss-Jordan elimination. Perform $-2R1 + R2 \to R2$:

$$\begin{bmatrix} 1 & -2 & \vdots & 1 & 0 \\ 0 & 7 & \vdots & -2 & 1 \end{bmatrix}$$

Perform $\frac{1}{7}R2 \to R2$:

$$\begin{bmatrix} 1 & -2 & \vdots & 1 & 0 \\ 0 & 1 & \vdots & -\frac{2}{7} & \frac{1}{7} \end{bmatrix}$$

Perform $2R2 + R1 \to R1$:

$$\begin{bmatrix} 1 & 0 & \vdots & \frac{3}{7} & \frac{2}{7} \\ 0 & 1 & \vdots & -\frac{2}{7} & \frac{1}{7} \end{bmatrix}$$

$\Rightarrow P = \begin{bmatrix} \frac{3}{7} & \frac{2}{7} \\ -\frac{2}{7} & \frac{1}{7} \end{bmatrix}$.

Let's check that $P\begin{bmatrix}4\\15\end{bmatrix} = \begin{bmatrix}6\\1\end{bmatrix}$:

$$\begin{bmatrix} \frac{3}{7} & \frac{2}{7} \\ -\frac{2}{7} & \frac{1}{7} \end{bmatrix}\begin{bmatrix}4\\15\end{bmatrix} = \begin{bmatrix} \frac{12}{7} + \frac{30}{7} \\ -\frac{8}{7} + \frac{15}{7} \end{bmatrix} = \begin{bmatrix} \frac{42}{7} \\ \frac{7}{7} \end{bmatrix} = \begin{bmatrix}6\\1\end{bmatrix}$$

Let's do another example.

Find the transition matrix from B to B' where $B = \{(1,3), (-2,-2)\}$ and $B' = \{(-12,0), (-4,4)\}$.

Form the augmented matrix $[B' \vdots B]$.

$$\begin{bmatrix} -12 & -4 & \vdots & 1 & -2 \\ 0 & 4 & \vdots & 3 & -2 \end{bmatrix}$$

Let's do Gauss-Jordan elimination.

Perform $-\frac{1}{12}R1 \to R1$:

$$\begin{bmatrix} 1 & \frac{1}{3} & \vdots & -\frac{1}{12} & \frac{1}{6} \\ 0 & 4 & \vdots & 3 & -2 \end{bmatrix}$$

Perform $\frac{1}{4}R2 \to R2$:

$$\begin{bmatrix} 1 & \frac{1}{3} & \vdots & -\frac{1}{12} & \frac{1}{6} \\ 0 & 1 & \vdots & \frac{3}{4} & -\frac{1}{2} \end{bmatrix}$$

Perform $-\frac{1}{3}R2 + R1 \to R1$:

$$\begin{bmatrix} 1 & 0 & \vdots & -\frac{1}{3} & \frac{1}{3} \\ 0 & 1 & \vdots & \frac{3}{4} & -\frac{1}{2} \end{bmatrix}$$

$$\Rightarrow P = \begin{bmatrix} -\frac{1}{3} & \frac{1}{3} \\ \frac{3}{4} & -\frac{1}{2} \end{bmatrix}.$$

Suppose $[x]_B = \begin{bmatrix}-1\\5\end{bmatrix}$. Find $[x]_{B'}$.

$$P[x]_B = P\begin{bmatrix}-1\\5\end{bmatrix} = \begin{bmatrix} -\frac{1}{3} & \frac{1}{3} \\ \frac{3}{4} & -\frac{1}{2} \end{bmatrix}\begin{bmatrix}-1\\5\end{bmatrix} = \begin{bmatrix} 2 \\ -\frac{13}{4} \end{bmatrix}$$

So $[x]_{B'} = P[x]_B = \begin{bmatrix} 2 \\ -\frac{13}{4} \end{bmatrix}$.

Let's check. $x = -1 \cdot (1,3) + 5 \cdot (-2,-2) = (-1,-3) + (-10,-10) = (-11,-13)$.

Now, calculate $2 \cdot (-12,0) - \frac{13}{4} \cdot (-4,4) = (-24,0) + (13,-13) = (-11,-13)$, the same result.

It's also possible to change the basis in the other direction, from B' to B. If P is the transition matrix from B to B', then P^{-1} is the transition matrix from B' to B.

In our example, if we want the transition matrix from B' to B, we just find the inverse of $P = \begin{bmatrix} -\frac{1}{3} & \frac{1}{3} \\ \frac{3}{4} & -\frac{1}{2} \end{bmatrix}$.

$$P^{-1} = \frac{1}{\det P} \begin{bmatrix} -\frac{1}{2} & -\frac{1}{3} \\ -\frac{3}{4} & -\frac{1}{3} \end{bmatrix} = \frac{1}{\left(-\frac{1}{12}\right)} \begin{bmatrix} -\frac{1}{2} & -\frac{1}{3} \\ -\frac{3}{4} & -\frac{1}{3} \end{bmatrix}$$

$$= -12 \begin{bmatrix} -\frac{1}{2} & -\frac{1}{3} \\ -\frac{3}{4} & -\frac{1}{3} \end{bmatrix}$$

$$= \begin{bmatrix} 6 & 4 \\ 9 & 4 \end{bmatrix}$$

PROBLEM SET: COORDINATES AND CHANGE OF BASIS

1. Find the coordinate matrix of x relative to the basis B'.

   $x = (5, 3)$

   $B' = \{(2, 4), (1, 3)\}$

2. Find the transition matrix from B to B' where $B = \{(1, 0), (0, 1)\}$, $B' = \{(2, 4), (1, 3)\}$. Then verify that $[x]_{B'} = P[x]_B$, for $x = (5, 3)$, using the answer from problem 1.

3. Find the transition matrix from B to B' where $B = \{(2, -2), (6, 3)\}$, $B' = \{(1, 1), (32, 31)\}$. Verify that $[x]_{B'} = P[x]_B$ for the vector x whose coordinate matrix relative to B is given by $[x]_B = \begin{bmatrix} 2 \\ -1 \end{bmatrix}$.

SOLUTION SET: COORDINATES AND CHANGE OF BASIS

1. We need to find scalars c_1 and c_2 such that $(5,3) = c_1(2,4) + c_2(1,3)$.

   $\Rightarrow \quad (5,3) = (2c_1 + c_2, 4c_1 + 3c_2)$

   $\Rightarrow \quad 2c_1 + c_2 = 5$

   $\qquad 4c_1 + 3c_2 = 3$

   $\Rightarrow \quad c_2 = -7$ and $c_1 = 6$.

   $\Rightarrow \quad [x]_{B'} = \begin{bmatrix} 6 \\ -7 \end{bmatrix}.$

2. Form $[B' \vdots B]$ and perform Gauss-Jordan elimination to get $[I \vdots P]$.

   $$\begin{bmatrix} 2 & 1 & \vdots & 1 & 0 \\ 4 & 3 & \vdots & 0 & 1 \end{bmatrix}$$

   The final result is

   $$\begin{bmatrix} 1 & 0 & \vdots & \frac{3}{2} & -\frac{1}{2} \\ 0 & 1 & \vdots & -2 & 1 \end{bmatrix}$$

   $\Rightarrow \quad P = \begin{bmatrix} \frac{3}{2} & -\frac{1}{2} \\ -2 & 1 \end{bmatrix}$

   Now, let's verify that $[x]_{B'} = P[x]_B$ for $x = (5,3)$.

   $$P[x]_B = \begin{bmatrix} \frac{3}{2} & -\frac{1}{2} \\ -2 & 1 \end{bmatrix} \begin{bmatrix} 5 \\ 3 \end{bmatrix} = \begin{bmatrix} 6 \\ -7 \end{bmatrix}$$

   $\qquad\qquad\qquad\qquad\qquad\qquad = [x]_{B'}$ from problem 1.

3. Form $[B' \vdots B]$ and perform Gauss-Jordan elimination to get $[I \vdots P]$.

   $$\begin{bmatrix} 1 & 32 & \vdots & 2 & 6 \\ 1 & 31 & \vdots & -2 & 3 \end{bmatrix}$$

   The final result is

   $$\begin{bmatrix} 1 & 0 & \vdots & -126 & -90 \\ 0 & 1 & \vdots & 4 & 3 \end{bmatrix}$$

   $\Rightarrow \quad P = \begin{bmatrix} -126 & -90 \\ 4 & 3 \end{bmatrix}$

   Now, let's verify that $[x]_{B'} = P[x]_B$ for the vector x whose coordinate matrix relative to B is given by $[x]_B = \begin{bmatrix} 2 \\ -1 \end{bmatrix}.$

$$P[x]_B = \begin{bmatrix} -126 & -90 \\ 4 & 3 \end{bmatrix} \begin{bmatrix} 2 \\ -1 \end{bmatrix} = \begin{bmatrix} -162 \\ 5 \end{bmatrix}$$

Since $[x]_B = \begin{bmatrix} 2 \\ -1 \end{bmatrix}$, $x = 2 \cdot (2, -2) + (-1) \cdot (6, 3) = (4, -4) + (-6, -3) = (-2, -7)$.

To verify that $[x]_{B'} = P[x]_B$ for our vector x, we need to show that $[x]_{B'} = \begin{bmatrix} -162 \\ 5 \end{bmatrix}$. That is, we need to show that the coordinates of x relative to the basis $B' = \{(1,1), (32, 31)\}$ are -162 and 5. So we need to check that $x = -162 \cdot (1, 1) + 5 \cdot (32, 31)$.

$$-162 \cdot (1, 1) + 5 \cdot (32, 31) = (-162, -162) + (160, 155) = (-2, -7) = x$$

So $[x]_{B'} = P[x]_B$ for our vector $x = (-2, -7)$. Of course, $[x]_{B'} = P[x]_B$ for every vector x in $\mathbb{R}^2$.

SUMMARY: BASIS AND DIMENSION

- If $S = \{v_1, \ldots, v_k\}$ both spans V and is linearly independent, then S is said to be a **basis** for V.

- For $\mathbb{R}^n$, in general, the standard basis is $\{(1,0,\ldots,0), (0,1,0,\ldots,0), \ldots, (0,0,\ldots,1)\}$, where there are n vectors in the set.

- The standard basis for P_n, the set of all polynomials of degree less than or equal to n, is $\{1, x, x^2, \ldots, x^n\}$.

- It turns out that, if $\{v_1, \ldots, v_k\}$ is a basis for V, not only can every vector in V be represented as a linear combination of $v_1, \ldots, v_k$, but the representation is unique.

- If V has a basis consisting of n vectors, then the **dimension** of V is n. The dimension of V is unambiguous because every other basis for V has n vectors.

- If x is an arbitrary vector in V, and B is a basis for V, then x can be written as a linear combination of the vectors in B. So if $B = \{v_1, \ldots, v_n\}$, then $x = c_1 v_1 + \cdots + c_n v_n$ for some scalars $c_1, \ldots, c_n$. The scalars $c_1, \ldots, c_n$ are called the **coordinates** of x relative to the basis B.

- We can form a column matrix consisting of the coordinates of x relative to B as follows:

$$\begin{bmatrix} c_1 \\ c_2 \\ \vdots \\ c_n \end{bmatrix}$$

This is called the **coordinate matrix** of x relative to B and is denoted $[x]_B$.

- There is a matrix P, called the **transition matrix** from B to B', such that $P[x]_B = [x]_{B'}$. If we're given the coordinate matrix for x relative to B, we can simply multiply by the transition matrix P and the result will be the coordinate matrix for x relative to B'.

- There is a procedure for finding the transition matrix P. Form the augmented matrix $[B' \vdots B]$ and perform Gauss-Jordan elimination to get $[I_n \vdots P]$.

- If P is the transition matrix from B to B', then P^{-1} is the transition matrix from B' to B.

CONCLUSION

Congratulations on completing the Linear Algebra book! Here is a review of what we have covered in this course:

- **Gaussian Elimination**
- **Vectors**
- **Matrix Algebra**
- **Determinants**
- **Vector Spaces**
- **Subspaces**
- **Span and Linear Independence**
- **Basis and Dimension**

I hope this book has been useful to you, and I wish you the best in your career and future endeavors. If you feel that you've benefitted from this course, I'd really appreciate it if you wrote a short review for the book.

Be sure to get the companion online course Linear Algebra for Beginners here: https://www.onlinemathtraining.com/linear-algebra/. For more online courses, visit: http://www.onlinemathtraining.com/.

Thank you, again!

Richard Han

INDEX

additive identity, 51, 60
additive inverse, 51, 60
augmented matrix, 20
basis, 121, 136
cofactor, 72, 78
cofactor expansion, 71, 78
coordinate matrix, 129, 136
coordinates, 129, 136
dimension, 123, 136
elementary row operations, 20
Gaussian elimination, 10
Gauss-Jordan elimination, 62, 70
inverse, 61, 70
linear combination, 30, 42
linearly dependent, 35, 42, 112, 120
linearly independent, 35, 42, 112, 120

minor, 72, 78
multiplicative identity, 52, 60
nontrivial subspace, 102, 108
row echelon form, 13
span of S, 111, 120
spans, 109, 120
square matrix, 45
standard basis, 121
subspace, 101, 108
transition matrix, 130, 136
transpose, 57, 60
trivial subspaces, 102, 108
vector space, 86, 100
weights, 30, 42
zero matrix, 51
zero subspace, 102, 108

CPSIA information can be obtained
at www.ICGtesting.com
Printed in the USA
LVHW060332311219
642165LV00018B/1722/P